INTRODUCING LIFE INTO PHYSICS

Thomas A. Behling

Edited by:
Sharon Nolan-Rosager

Copyright 1988

ISBN 0-9626256-0-4

Dedicated to:

Karlakay Christopherson –

Whose benevolence allowed this work to be drafted shortly before her untimely death.

LETTER OF TRANSMITTAL

Dear Reader:

You will find the following work unorthodox in style. It reads much like a long letter of transmittal. Its logic patterns purposefully repetitive; at times profound; at times childlike inanity; and always disruptive.

I suppose one has to possess a certain amount of vanity to try to have his theory recognized – so if you seek that human flaw, you will surely find it, and while it may be vainly said the work's source is altruism, perhaps the greater truths are: it burned inside me; and I had to rid myself of it. Spent now, I know no other way to preserve your dignity than to say: this dialogue is yours if you so choose to receive it.

In faith, hope, and love;

Tom Behling

Contents

PROLOGUE i
THE UNITIVE EQUATION................................... 1
  1.0 INTRODUCTION.. 1
  2.0 EQUATION.. 3
      2.1. Postulations 3
          2.1.1. ENERGY AND MASS ARE THE SAME
                     SUBSTANCE................................... 3
          2.1.2. INTERACTION OF SUBSTANCE   3
          2.1.3. SPATIAL OCCUPATION 3
          2.1.4. MOTION AND CHRONOLOGY............... 3
          2.1.5. NATURAL LAW ABSOLUTE.................. 4
      2.2. Equation Derivation... 4
      2.3. Features of Equation....................................... 5
          2.3.1. PARAMETRIC-ORTHOGONAL GROWTH .. 5
          2.3.2. n ROOTS... 5
          2.3.3. MASS–ENERGY CHARACTERISTICS....... 5
          2.3.4. UNITIVE CHARACTERISTICS................. 5
      2.4. Title of Equation.. 6
      2.5. Motivation... 7
      2.6. Einstein's Golden Calf.................................... 8
      2.7. Unitive Theory's Approach 8
      2.8. Show and Tell 10
  3.0 OUR EARTH'S PARAGENESIS............................ 13
      3.1. Ancestral Time... 13
      3.2. Solution.. 13
          3.2.1. ASSUMPTIONS 13
          3.2.2. DISCUSSION 14
          3.2.3. SOLUTION ATTEMPT 14
          3.2.4. PRELIMINARY CONCLUSION 17
      3.3. From the Top Down 18
          3.3.1. DATA 18
          3.3.2. ASSUMPTIONS 18
          3.3.3. REGRESSION.................................... 19
              3.3.3.1 First Premise 19
              3.3.3.2 Second Premise......................... 20
              3.3.3.3 Redefinition of Inertia.................... 21
              3.3.3.4 Reflection Problems..................... 22
              3.3.3.5 Gravity...................................... 22
              3.3.3.6 Tides 31
              3.3.3.7 Gravity Energy? 36

- 3.3.3.8 The Theory of Relativity vs Wave Mechanics. ... 39
- 3.3.3.9 Uniting Historical Physics ... 41
- 3.3.3.10 The Speed of Light ... 64
- 3.3.3.11 The Future ... 65
- 3.3.3.12 The Sun Continued ... 73
- 3.3.4 GENESIS ... 76
  - 3.3.4.1 Introduction ... 76
  - 3.3.4.2 Day One ... 77
  - 3.3.4.3 Day Two ... 78
  - 3.3.4.4 Day Three ... 78
  - 3.3.4.5 Day Four ... 80
  - 3.3.4.6 Days Five and Six ... 81
- 3.3.5 OTHER QUESTIONS ... 82
  - 3.3.5.1 Sun Spiral ... 83
  - 3.3.5.2 Similar pre-Cambrian Rock Types ... 83
  - 3.3.5.3 Ocean to Land H_2O Transfer ... 83
  - 3.3.5.4 Man's Creation ... 83
  - 3.3.5.5 Atomic Structure ... 84
  - 3.3.5.6 Speed of Light ... 85
  - 3.3.5.7 de Broglie's Equation ... 85
  - 3.3.5.8 Show and Tell ... 85
  - 3.3.5.9 Partially Solved Problems ... 89
- 3.4. Quantifying Life ... 94
  - 3.4.1. MATHEMATIC PREMISE ... 94
- 4.0 CLOSING ... 99
- 5.0 THE FOUR SEASONS ... 101
- 6.0 ACKNOWLEDGEMENT ... 105
- 7.0 EPILOG ... 107
- ADDENDUM Dual Function Equation ... 109

List of Illustrations

Figure 3–1a, –1b, –1c : Paragenesis of the Earth from the Sun ... 16
Figure 3–2 : Mass distribution of Earth ... 24
Table 3–1: Comparison of "g" at the Earth's Surface ... 30
Figure 3–3: Iso-distance traces from the Moon to the Earth ... 31
Table 3–2: Ratios of planets' charge with number of satellites ... 58
Figure 3–4: Suggested Experiment ... 62
Table 3–3: Comparison of "a_o" ... 88
Table 3–4: G_p quantumized to a constant ... 91

PROLOGUE

You, the reader, must thresh out the logic patterns within this work. I do not find it becoming to shepherd to your destiny, and therein your dignity. Nor do I care to become your scorn due to your intolerance to peruse change. Even so, this preface should not be avoided if it enhances your threshing.

The original work written in 1978 is an addendum to the material immediately following. This previous work is evolutionary in nature. Definitive continuity between micro and macro mass energy systems is brought about by introducing two new variables *inertial potential* and *entropic potential;* however, the concept that *mass and energy are separate entities* is retained, as well as, that *free gravity flux is tied to the mass function* (all part of my educational programing).

The new portion, immediately following, began as a problem utilizing the original equation to explain why our ancient biblical ancestors "apparently" lived longer than we. What happened was the entropy of my pen would not cease, and after so many words spent, I realized I had reshaped my educational programing to inchoately incorporate life processes into physical law; furthermore, all *physical* substances have life. (This is not a godless approach to life, but commingles the readers *metaphysics* with his molecules. Perhaps there is meaning to life after death through resurrection of the body; as well as, life returns to the ground from whence it came.)

Ponder the following if you will:

Our sun, our earth and you at the time of your birth each had different *scalar* masses than today. (The amount of mass of each of these objects is different now, as well as, each object's molecular identity.) Where should I place my mathematical authority to engulf this dichotomy? Should I assume that each *existence* is *scalar* but has the ability to transcend both time and variable mass commingulings? Thus, giving each *existence* vectorship - a *scalar* with magnitude and direction? Thus, destiny? Or should I assume that each *mass* is a *scalar* and that each new *entity* (resulting object) takes on a different *identity* with each mass transition? In this latter assumption how can you remember your childhood's suns and earths if you are a different entity, or assess which childhood hopes you have accomplished?

i

Remember how much your toe hurt when you stubbed it? And yet your toe still remains with you; even though its molecules that hurt so much have since left you. If you credit your hope's successes to your *physical* ability to assimilate food energy and then *metaphysically* direct it toward the future, are your failures the result of your body physically storing this energy for its future dreams? By what mathematical system do we approach these *physical-metaphysical* comminglings?

This new work invests the former work found in the addendum to approach these comminglings. Neither the works of the Greeks, Newton, Einstein or Wave Mechanics promoters were left unscathed by this reshaping process. The universe's physical existence theoretically resides in one substance — gravity flux, which for some *metaphysical* reason has the ability to interact with similar flux in a state of net-cancellation — inertializing both flux fields — to form mass.

Mass then becomes those net-cancellations which have to be destroyed to utilize the flux energy within. Motion, as a phenomenon, is fundamental for this destruction, and is also fundamental for stimulating other new net-cancellations. All objects have <u>external</u> gravity flux fields due to the motion or resonance within their constituents. These residual flux fields can interact with other residual flux fields emanating from other bodies to form higher order resonances.

By visually recognizing that an object's gravity flux has the *metaphysical* ability to cause interaction with other gravity flux, spontaneity can be explained as a response-to-gradient phenomenon, and also I can treat you with dignity which preserves our *metaphysical* interactions.

In short, I concluded my assumption *"that gravity flux net-cancellations are the source of all physical entities";* however, not without much metaphysical struggle was this continuity found; to continue this struggle is yours. If you find some of the postulations and equations too stuffy, abstract or presumptuous, just keep on reading, you will always have them to come back to, and your evenings to shout thou blasphemer; thou fool.

If you wish to follow the chronology of this theory's development, start with the 1978 work.

To the youth:

You should not be disturbed if you do not fully comprehend this work at this time, but not so ironically the future of this work resides in you. As you further your academic education, you will want to compare this work to find the continuity between it and the established theories taught in your institutions of higher learning. It is not the purpose of universities to promote theories as this one for it has not yet withstood the test of time; rather, their purpose is to promote established theory to cause research to establish newer theories. Be good students and learn the established theories before you use this work to effect changes on the older works. I do not find the older theories so much wrong, but rather deficient in depth of logic. The following are examples:

1. Energy has been defined as "the capability to do work." This "capability" gives energy a metaphysical property which the older theories have not addressed properly because with "capability" comes "incapability"; thus, energy should equal "the capability to do work" plus the "the capability to create inertia," or resistance to work. This definition of energy is not yet complete because "capability" is not a substance. It is only a metaphysical property; therefore, a more complete definition of energy is – energy is a "substance" which occupies a space-time relationship with the "capability" to create inertia (fixed energy) plus entropy (free energy). More simply – E = (inertia + entropy). (The amount of work done may <u>not</u> equal the amount of entropy, but the net difference will show up as inertia.) We cannot allow the classical definition of energy to dominate over the "substance" energy. Rather it seems more prudent to determine how many ways energy can be utilized and then broaden our definition to include all of these utilizations.

Two simple examples:

    a. Suppose you push on your front door to prevent an intruder from coming in. If the door does not move, you are said to have done no work. On what did you expend your internal energy? Answer – Creating a directional state of inertia on your front door.

    b. Suppose you attempt to pick up a (1) kilogram ball, but you only apply enough energy to nullify its weight downward. On what did you expend your energy? Answer – Latent heat or overcoming the inertia of the ball's and earth's gravity couple. Any additional energy upward will cause the ball and earth to undergo a physical change of state. No different than a molecule sublimating away from a solid.

If we can utilize energy for the above purposes, why are they not included in the definition of energy?

2. You cannot see free energy - you may see a car speeding down the street but you cannot see its free or kinetic energy. It takes metaphysical experience to understand that the car may hurt you. This plays an important role in how we place our faith in what we cannot see but can experience. It also divides and combines the physical and the metaphysical realms.

3. This lack of visibility of free energy plays an important role in understanding the phenomena of inertia (fixed or stabilized energy) and entropy (free or kinetic energy). If we accelerate an object, we impart free energy into the object to rob it of its inertia, but the free energy imparts an increase in "stability" to the object in the direction of its motion. "Stability" then may result from the absence of free energy or the use of free energy. In the past, inertia represented both types of this "stability," and entropy represented "disorder" or "instability" caused by the motion. I concluded that redefinitions of these terms were in order because motion of an object causes both inertia and entropy.

4. Mass is a phenomena, as well as inertia, but this theory contends they are synonymous and the same. Only our point of view causes us to perceive them differently. We cannot see the free energy of the gravity couples that hold the earth and ourselves together, but the sun has to move these couples as it moves the earth and ourselves through its orbit; thus, not only the mass of the earth, and of ourselves, but also these free energy couples represent mass to the sun. This lack of ability to see free energy has caused us to believe gravity is only a force. It is a more complete concept to say a force is our incomplete experience of applied energy; in this way we can then represent all phenomenon to one substance "gravity" or the energy of existence.

5. In the past, physicists dealt with the mass of an object to study the object's behavior under certain conditions; however, objects have more complete properties than just mass. You as an object recycle your mass periodically by ingesting food and excreting refuse; thus, at some future time "you" will be a different mass than you are today. Your "existence" or "life" vector has the "ability" (a metaphysical property) to sustain its existence by altering its mass characteristics (all objects have

existence and therefore interact with their environment to sustain themselves just as you do but to a less complex extent).

6. When we pick up an object from the earth much of the energy spent is in separating the binary earth – object gravity couple. In outer space far from the earth, this is not the case, most of the energy will be transferred to the same object; thus it is that the same object can exhibit variable gravity characteristics. As you will see, the phase angle in the equation should account for this change in gravity state.

7. If you understand conservation of energy, then you should also understand that an "object" can never be more or less than itself, or there is a state of "oneness" or "unity" about an "object" during any instant of time.

As author of this work, I suggest to the youth that your *first* reading of this material be a rapid one with little time spent wrestling with its purpose or postulations. In this way, you will library the different subject matters addressed so that you may return to them at a later date. I fear that a tedious or contemplative first reading may cause you to never finish it, and this would be self defeating.

THE UNITIVE EQUATION

1.0 INTRODUCTION

The author is a graduate of Montana Tech in the fields of Mining & Geological Engineering, 1974. Although I find Physics interesting, it is not my choice of professional careers. This presents one of those dichotomies in life which causes me to rely on the three universal truths; faith, hope & love.

My reason for this position is:

I believe that I have formulated an equation that does indeed combine Newtonian, Relativistic, Thermodynamic, and Quantum Physics into one, but because I am not a member of the scientific community which promotes such findings, the truth of the equation may never be received or understood. I can provide the understanding which caused its formulation to only those who are willing to listen. So far, no physicist has given me reason to rescind any of its postulations. It also provides more reasonable explanations for those physical phenomena which are not well explained by any of the systems above.

2.0 EQUATION

2.1. Postulations: (Postulations are always stuffy - so wade through them and then go on.)
This Unitive or Dual Function equation was derived from and also concluded the following concepts:

2.1.1. ENERGY AND MASS ARE THE SAME SUBSTANCE:
That energy and mass are the same substance, but differ in appearance due to variations in freedoms of spatial occupation; thus, not only both *E* and *m* typify what resonant form this substance will appear as, but also *E*'s and *m*'s environment police what resonant form will prevail;

2.1.2. INTERACTION OF SUBSTANCE:
That this substance does have the ever present ability to interact with substance of like character, and is confined to do so. More simply, mass and energy are expressions of confinement of this substance by two different processes. Mass/inertia results from net-cancellations of gravity flux (such as two equal "force" vectors in direct opposition). Field energy/entropy result from permeation of void by this substance interacting with like substance whose freedoms to do so are orthogonal to those net-cancellations which create mass/inertia. Orthogonal decay and growth of linear net-cancellations is the primary parametric format of this equation;

2.1.3. SPATIAL OCCUPATION:
Permeation of space by field energies radiating outward from mass, prevents void, i.e., a non-ethereal interpretation of our universe's interior is contradictory to the gravity that brings order to the heavens; therefore, a type of firmament exists throughout space, even though it is not readily characterized on a mass-energy mentality. (Void is an entity in this theory.) *Empirically by visual inspection,* this phenomenon of interaction of gravity flux brings about inertia to our galactic and solar systems; thus, I postulate gravity net-cancellations are complete energy vectors permeating space;

2.1.4. MOTION AND CHRONOLOGY:
That motion cannot exist without chronology; time warps exist only as an inner freedom and are not communicable but

fictitious. One outcome of this equation is that we no longer need the Theory of Relativity to explain "apparent" time warps. This new theory explains such phenomena from a transcended reference frame (a position where the sensitivities of inertial order and dynamic order are simultaneously examined as a unit.) Time is said to be absolute or unitive (i.e., all points interior to the universe are of the same time). (In the 1978 work, I considered such reference frames as inertial due to my educational programing; however, because Newtonian Laws become relativistic with transcendence, I concluded an inertial reference frame places all sensitivities (the mathematical authority) at the point of inertia, where as a transcended frame of reference places this mathematical authority into said problem's environment.) [Note: I believe Einstein made a false assumption when he concluded that the velocity of light is constant, and then extended this assumption to say that time is dependent on motion. Motion couples substance (the corporeal) with time and distance (the meta-corporeal); substance and time are not transmutable. A particle cannot gain mass just because its velocity vector is changing; this violates the law of energy conservation, because non substantive vectors cannot add substance (mass) to a particle, rather, acceleration is an indicator of what mass transformation is taking place between an object and its environment. Einstein's equation E = (a constant) (mass) is a first order equation when viewed from a substantive vector reference.] This transcended reference frame preserves an object's metaphysical nature by allowing it to exist while suffering both *internal* and *external* inertia-entropy transitions.

2.1.5. NATURAL LAW ABSOLUTE:
That natural law is absolute, perceivable, and therefore, upon investigation—understandable, but not changeable upon perception. Our minds are subordinate to those processes that caused their existence. Formulations, rationale and theology do exist, but only depict natural law.

2.2. Equation Derivation:
Getting away from the abstract, this equation states that in any mass - energy resonant system (an object or objects) — the total energy — is a function of the sum of the energy tied up as *mass/inertia* plus field *energy/entropy*.

$E = mc^2$ — (Suggests linear exchange one for another.)
$E = h\nu *$

Eq. 2.2.1. $E_t = mc^2 \ (1)^n = mc^2 \ (cos^2\theta + sin^2\theta)^n$ **

Eq. 2.2.2. $E_t = mc^2 \ cos^2\theta + h\nu sin^2\theta$ — (Orthogonal parametric exchange.)

* – $h\nu$ – figuratively represents kinetic energy or fluxed energy free to interact with like flux outside E_t's mass; as well as quanta (objects smaller than the object(s) E_t represents). Complex growth and decay can be facilitated by grouping or integrating several E_t's with the same or slightly different phase angles.

** This parametric format of (1) is the incorporeal (life or spirit) vector between the corporeal and meta-corporeal vectors (more on this later).

2.3. Features of Equation:

2.3.1. PARAMETRIC-ORTHOGONAL GROWTH:
The Equation's parametric-orthogonal growth and decay of energy forms, orthogonal to linear net-cancellations of vectors reflect the greatest freedoms for growth and decay;

2.3.2. n ROOTS:
The Equation has *n* roots which is believed to be important when discussing variations in freedoms for motion, ability for growth and decay, and ability for life properties ;

2.3.3. MASS–ENERGY CHARACTERISTICS:
The Equation never neglects to address either mass or flux energy; and

2.3.4. UNITIVE CHARACTERISTICS:
The Equation provides *unitive* thought for mass-energy characteristics for all energy transforms from photons to the largest of bodies — be they galaxies or "black holes."

Note: Consistent with this theory — "black holes" exist as large inertial zones due to the aggregate impingement of stellar gravity energies; and therefore, they are wave-front phenomena which can appear, diffuse, and reappear at enormous velocities, even though their existence, motions and characteristics are caused by much slower moving mass objects. This theory depicts "black-holes" as the macro of those net-cancellations (lack of freedoms) of field energy vectors which similarly create the micro. In this way, the theory avails room for [[(conservative system) within system] within system] interacting unitively with one another to form a universe whose curvature is maintained by linear net-cancellations acting radially outward/into each constituent.

2.4. Title of Equation:

A word about the title of this equation:
Originally, its title was the Dual Function Equation; however, as of late, I feel the title Unitive Equation is more descriptive, because after seven years of reflection, nothing has adversely disturbed me as to its positions when relative thought is abandoned for it transcended principles. (The Catholic definition of *unitive* as the highest level of mental prayer can be found below.)

> ***Prayer, Mental.*** The raising of the mind and heart to God without using words. Mental prayer can be of many types, and the various stages of the supernatural life are frequently distinguished by the type of mental prayer which characterizes them.
> ***Meditation or discursive mental prayer*** is that in which considerations and reasonings on a religious truth predominate over pious affections (e.g., acts of love) toward God. The truth that is the object of meditation may be a mystery of faith (e.g., the Trinity), an event in the life of Christ (e.g., raising Lazarus), an event in the life of the saints (e.g., Saint Peter's denial of Christ), a text from Scripture, the liturgy, or from the writings of the saints. This type of prayer gives one a deeper appreciation of the truth in question and leads to an application of that truth in one's life.

This type of prayer is characteristic of the *purgative way*.

Affective prayer is that in which one who has frequently meditated on a truth and has considered it under many aspects can, by merely recalling the truth, occupy himself with devout affections that is, with acts of will (e.g., acts of love and acts of hope). In this type of prayer, the affections predominate over reasoning. Such prayer characterizes the *illuminative way*.

Contemplation is a simple and affective gaze upon God or some divine truth. In this type of prayer, the mind simply looks upon a truth and contemplates it with admiration and love. Without the reasonings that mark discursive mental prayer or the multiplicity of affections which mark affective prayer, contemplation is a simple and loving gaze upon truth. This type of prayer characterizes the *unitive way*.

2.5. Motivation:

I realize that many scientists avoid intertwining the physical world they investigate with a spiritual text, such as the Bible, which at times appears quite allegorical and not very empirical. After all, it is rather peculiar that we as scientists seek truth, and at the same time, the Bible indicates the three greatest truths are faith, hope, and love which are not empirically provable. Metaphysically, these three motivating truths impinge us into/onto the material world around us to seek omniscience, whose incomplete position is the aggregate of opinion's yeas and nays. As it should be.

After finishing the addendum in 1978, I did in fact, try to introduce it to the scientific community, but for the most part its ears are deaf. Like "a voice crying in the wilderness," the precepts of the equation have yet been received. I suppose I could be (and probably to some, should be!) out chasing women or some similar gregarious activity, rather than trying to remove the "golden calves" of modern physics. Because both the new and original work came to me through much prayer and quiet reflection, and perhaps most importantly, through those seekers who preceded me, I feel I owe it to their presence to persist in an exchange of dialogue.

7

2.6. Einstein's Golden Calf:

Although this same equation does not violate the equivalence found in Einstein's – $E = mc^2$ equation, I do believe Einstein put aside divine guidance for materialism when he assumed electromotive transfers could take place without an ether. This "golden calf" was apparently "proven" with the results of the Michaelson-Morley experiment. The conflict here is: it was presumed that the ether they sought was separate from the earth – much like a ball passing through a liquid. (The Unitive Theory does not, but maintains that this ether they sought is inherently the earth's gravity field interacting radially like the sun's.) Einstein's postulation neglected to account for the empirical data then at hand, i.e., that the gravity fields which permeate space were also in their experiment, this violation is borne out in Einstein's higher postulations i.e., that dense gravity fields *can* affect a photon transfer. How can dense "void" affect an energy transfer? (When Einstein postulated the velocity of light quasi-constant, he presumptuously assumed that gravity flux was not interacting with the photon's gravity flux and created another mathematical system to analyze resonant mass-energy. Once he effected this system, mathematicians since have justified its presence. In short, he gave metaphysical life to *c*.)

2.7. Unitive Theory's Approach:

It is much more realistic to confront this problem by identifying that gravity fields permeate space and that interactions of these fields with "electromotive" transfers do affect the velocity and direction of these transfers; thus, the velocity of light is not constant but dependent on its environmental freedom to interact with surrounding substantive vectors (medias).

With the Unitive Theory such exchanges become transcending events, its equation systematizes light so that it has both mass and field energy characteristics, i.e., corpuscular variations in a photon transfer relegate variations in the density of the photon's gravity flux or freedom of energy permeation, which is in fact, the theory's definition of mass versus field energy.

In explanation of this theory, I belittle no scientists past or present and that is especially true with Albert Einstein, a dear man who spent his entire life illuminating the physical world to mankind. After all, the Unitive Equation is an expansion of his Relativistic Equation (not the first time in history a Christian expanded a Jewish tenant to further illuminate God to mankind.) I do believe that Einstein was a victim of what I refer to as "Grecian Logic." The ancient Greeks deified gods from material objects, and their thinkers left us with the idea that the universe is built of the infinitesimal, atoms. This trend in mankind's logic causes scientists to promote the micro findings in the laboratory to explain the macro events of the universe.

The Law of Subordination:

The Unitive Equation does in fact the opposite; and therefore, God becomes a presence once again as ultimate Being. Why? When I originally began its formulation, I too started with the infinitesimal, i.e., the photon, but treated it as a parametric unity of mass and field energy constantly changing (recalling that corpuscular variations, relegate variations in flux freedoms.) In the 1978 work, I repeatedly applied this parametric photon to account for light's emissive, momentive, refractive, diffractive, and reflective properties, and just as repeatedly, I was led to the conclusion – that a photon only exists where its environment allows it to exist. And as we look to the empirical data of our universe, do we not see the same event happening? The small objects therein are ordered by the large; therefore, I concluded the photon has to interact with its environment to exist and is subordinate to it. Based on this rationale, I had to knock the anthropomorphic stuffing out of Einstein's c. Once again, the change in c or any velocity vector may depict an object's state of mass change in a certain environment, but c is a non substantive vector and does not cause the mass change, the mass metamorphism is caused by addition or subtraction of gravity flux net-cancellations.

[Regression: Photons are always ordered by the mass they pass by or through. Empirically it can be shown that photons passing through a vacuum, in near proximity to an object of mass will be bent toward the mass (there may be those zones around decaying objects which will bend a photon away from it due to solar wind caused by the high potential for photons to escape from these areas). If we assume that this bending action is caused by the interaction of gravity fields, and if we then identify this same empirical phenomena happening between small objects orbiting about the more inertial of the heavens, we will conclude that the permeation of these spatial expanses by said gravity flux fields (in states of net-cancellation) will provide a complete energy vehicle for the photon to interact with. Conversely, we cannot assume our solar system and universe are held together by the interaction of gravity flux fields, and also assume that a photon has a gravity field, without assuming that an interaction is possible between both fields.

Also, inherent within the empirical data at hand is the more inertial an object, the greater the potential for its decay by radiation, provided that its environment allows that decay. Thus, solar emissions, radioactivity, and electron emissions etc., are not spontaneous random processes but are based on potentials and environmental freedoms. (An interesting point comes to mind here, if one flies a radioactive clock through the earth's gravity field in different directions – different emission counts will result per time – one does not warp time in this experiment but creates different freedoms for radioactive decay. Documenting this will again prove that universal time is unitive and not dependent on motion.)

Because of these two opposing characteristics of mass-energy, the author was led to the conclusion that two new variables were needed (new in that I have never heard of them before), i.e., **inertial potential** and **entropic potential** which are inversely related to inertia and entropy respectively (see the 1978 work).]

2.8. Show and Tell:

1. Given: $E_t = mc^2 \cos^2\theta + h\nu\sin^2\theta$:

   a. Show that at $\theta = 0°$ and $90°$ the equation is bounded by Einstein's and Planck's equations respectively from which it was derived.

   b. Show that E can be a standing or travelling wave function (see Addendum, pages 1 and 3).

2. Given: The speed of light is a finite value:

   a. Show that time can approach zero but never become zero.

   b. Convince yourself that if **motion** did not exist — time could not be measured.

3. Given: That time has both quantity (duration) and quality (object rapport to environment) covenants:

   a. Show that **time warps** are exclusive to time's quality covenant with you.

   b. Does Jupiter have to speed up because you suffered a **time warp**?

11

3.0 OUR EARTH'S PARAGENESIS

3.1. Ancestral Time:

During the past Easter season I was meditating on something in the Bible when I was struck with the awareness that the registered accounts of the ancients before the flood and shortly after depict extremely long lives. Moses, who is believed to have written the book of Genesis, describes those who lived 900 years plus. As a scientist, I found these life spans rather empirical, after all if someone lived 90 years or so, why should someone else record that he lived 900 years? I then rationalized that according to the Unitive Theory time is absolute or unitive and not relative. It then becomes apparent that a day and a year are events, that is one revolution of the earth and one orbit about the sun respectively.

3.2. Solution:

3.2.1. ASSUMPTIONS:

I attempted a solution to this problem with the following assumptions:

1. That we and the ancients lived approximately the same absolute time;

2. That because days and years are events, tenuous conflicts between geological time, biblical time and relativistic time are put aside for the chronology of these events. (The Geologic time table expresses the evolution of the earth in 4 to 5 billion years, the Bible depicts this evolution happened in six days, and the Theory of Relativity, in theory, could be utilized to compress this time of evolution to six minutes at extremely high velocities.);

3. That the earth and moon have represented a fixed mass system over its lifetime - except for the earth possibly entrapping the moon as a satellite (prior to the ancients), and also the earth's and moon's constant entrapment of solar energy and meteorites (most of which was prior to the ancients). This assumption does not imply fixed configuration of mass;

4. That due to the gyratory stability of both the sun and the earth rotating, the alignment of their axes has been roughly the same throughout their existence. (This statement does not preclude a constant-infinitesimal progression wherein the axial alignments of both masses are in a state of change, but rather; this statement recognizes the gyratory stability within each body.); and

5. That at one time the earth was too hot and/or cold to have seas, and at the time of the ancients (pre-flood era) seas may have been a minor feature, or today's seas were brought about by an evolutionary process.

3.2.2. DISCUSSION:

A host of sketchy geologic evidence would indicate that the earth has been expanding.

Some Examples:

Ancient shield rocks are more dense; so too the metallic epochs of gold through copper from pre-Cambrian to Mesozoic time might express a paragenetic compatibility to a decreasing density in the crust; the evolution of life forms, physically stable to their environment, might express a compatibility wherein less dense species of matter came into existence during this process of decreasing density; the incorporation of water into mineral structures within the crust has increased its volume, as well as physical expansion of shield and plutonic rock to form clastic sediments with voids; also, the moon's pull on the crust would cause it to expand or behave as a less dense material.

3.2.3. SOLUTION ATTEMPT:

In my first attempt to solve this problem using the Unitive Equation. I reduced the surface area of the earth to 30% of its present size, i.e., based on recollections of a former study wherein another geologist concluded that if you shift the continents around and match rock types you will create a sphere of smaller size, I removed the area of the oceans from our earth's present sphere of radius (r_1) to determine the original radius (r_o).

$$4\pi r_1^2 \times 30\% = 4\pi r_0^2,$$

$$r_0 = (0.30 \times r_1^2)^{1/2},$$

$$r_0 = 3490 km; \text{ when } r_1 = 6378 km.$$

Assuming both spheres have the same mass, the angular velocity increases to:

$$I_0 \omega_0 = I_1 \omega_1,$$

$$\omega_0 = (I_1 \omega_1) / (I_0),$$

$$\omega_0 = (2/5\ mr_1^2 \omega_1) / (2/5\ mr_0^2),$$

$$\omega_0 = r_1^2 \omega_1 / r_0^2.$$

Assume: $\omega_1 = (1)$, then

$$\omega_0 = 3.33 \omega_1.$$

This answer is off by a factor of 2 to 3 because the ancients lived about 7.5 to 10 times longer than we do in years. (We could attribute this to the premise that our ancestors actually lived 2 to 3 times longer in absolute time. This is not an unreasonable presumption because Noah lived ten generations after the flood until Abram's generation which would indicate Noah's longevity (950 years) was in part related to genetics. The earth was probably slowing down during these ten generations, but the absolute time would be experienced by both Noah and these ten generations. The author speculates that as the earth was slowing down due to the waters of the earth moving to the equator, the inertia-entropy ratio of the earth was changing which left its imprint on each generation's genes, producing shorter and shorter longevities.) My next attempt to calculate ω_0 for the earth was based on a nonlinear density gradient from surface to center. Using a density of approximately 2.65gr/cm³ at the surface failed to change I_o appreciably; therefore, I concluded that it was not *just* the expansion of the earth that would cause ω_1 of the earth to decrease. Because much of this theory came by evolution, I allowed the problem to proceed to see if I could better explain our origin. Ignoring momentarily the other planets in our solar system and employing the Unitive Equation's concept (see *Figure 3-1a*), I rotated the earth upward in solar latitude. The reason why I rotated the earth up in latitude was to keep the distance between the earth and the sun the same (I didn't want to burn up our ancient ancestors).

Figure 3–1a, –1b, –1c:

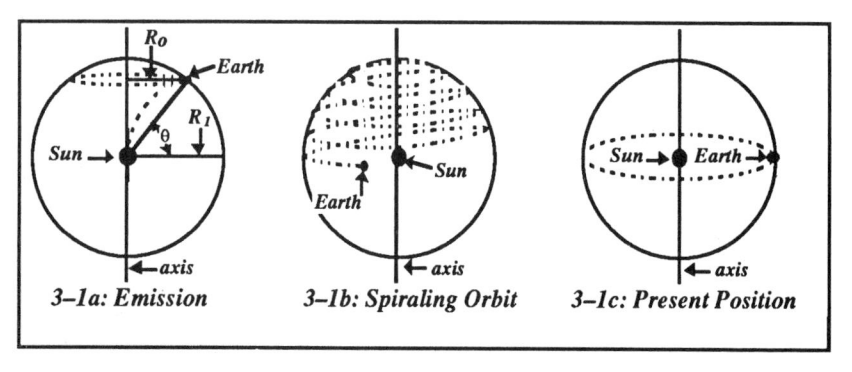

| 3–1a: Emission | 3–1b: Spiraling Orbit | 3–1c: Present Position |

Figure 3–1a,–1b, –1c: Showing paragenesis of the Earth from the Sun.

Referring to the former calculations where $\omega_o = 3.33\omega_1$, the corresponding angle *(θ)* in *Figure 3-1a.* would be:

$R_o = R_1 \cos\theta$,

$\cos\theta = R_o / R_1$.

By the former relationship of:

$\omega_o r_o^2 = \omega_1 r_1^2$,

$\omega_o = 3.33\omega_1$ for the earth's axial spin.

The orbital moment of inertia, $I = mR^2$; thus, to conserve the orbital angular momentum:

$$mR_I{}^2\Omega_I = mR_o{}^2\Omega_o,$$

and to maintain the corresponding relationship of 365 rev/orbit,

$$\Omega_o = 3.33\Omega_I,$$

$$R_o{}^2 = (R_I{}^2\Omega_I)/(\Omega_o), \text{ or}$$

$$R_o{}^2/R_I{}^2 = \Omega_I/\Omega_o,$$

$$\cos\theta = R_o/R_I = (\Omega_I/\Omega_o)^{1/2} = (1/3.33)^{1/2} = 0.55,$$

$$\cos\theta^{-1} = 57°.$$

The quantitative exactness of θ is based on the presumption that throughout this evolutionary process 365 rev/orbit was maintained; it needn't. Also, the shape of the solar sphere traced by the precession of the earth spiraling down in latitude could vary from a true sphere both horizontally and vertically to the sun's equator.

3.2.4. PRELIMINARY CONCLUSION:

By using this equation, I came to the preliminary conclusion that our solar system may have been created by a top-like paragenesis. (I believe it an endured concept that the planets of our solar system were formed near the sun's equatorial plane, either by acceleration of some of the sun's mass out into orbits or by equatorial rings condensing.) The concepts of the Unitive Equation depict that our planets could have been shot out from the sun's pole region and gradually spiraled down through the latitudes of the sun's spheres (see Figures 3-1b, and -1c.); furthermore, they could have been shot out all at once, or one at a time. The similarities and variations in axial tilts between our solar system's planets may depict this paragenesis; i.e., variation in our sun's spatial orientation upon ejection of our planets, may explain why our planets have similar but varying tilts.

3.3. From the Top Down:

How and why could this top-like paragenesis happen?

3.3.1. DATA:

Empirical data at hand:

1. The sun rotates;
2. The sun is in a state of mass decay;
3. All the planets in the sun's system rotate counter clock-wise the same as the sun;
4. All the planet's orbits are counter clockwise;
5. There is a trend towards faster axial rotation of the planets as the radius of orbit increases;
6. The total mechanical energy of the system appears fixed except for loss due to solar radiation; and
7. I am obviously not going to solve this problem with my hand-held programmable calculator, so I will outline the solution per the Unitive Equation for anyone who wishes to tackle it. It would be a good problem for a graduate study.

3.3.2. ASSUMPTIONS:

We first have to make some assumptions to get the problem started. Because the universe expresses a tenacious rigidity to maintain order by motion, let us assume that our neophyte sun prior to planet ejection was in fact in motion and that its motions were not very complex. For now, we won't assume that our sun was spiraling, spinning etc.; rather, it was just being ordered around by some larger mass aggregate. Let us further assume that it was in a state of mass growth. Why? The Unitive Equation predicts cyclical mass growth and decay during existence, much like our mother earth is presently growing due to the phenomenon of entrapment of smaller bodies. In reality, all we are saying is that our neophyte sun's motion caused it to pass through a zone of the universe made up of small bodies which it entrapped. But, some time thereafter, something different happened to our neophyte sun, its motions brought it to a zone of appreciably more-mass-density per cubic-light-year, and it had to grow up or get bounced back where it came from (reflect on that awhile).

Predictable by the logic of the Unitive Equation, our neophyte sun's energy (inertia-entropy) ratio must be sufficiently compatible to interact with this more-dense-zone-of-mass, or it will not become part of it. [Reflection is the physical phenomenon wherein a particle's energy (inertia-entropy) ratio will not interact with its media, and reflection is elastic to the extent that the particle's momentums before and after are equivalent. Reflection is evidence that the *large* determine the location of, or *order* the *small*.] Let's assume that our neophyte sun did in fact have sufficient energy ratio compatibility to interact with this dense mass zone. What would its behavior be? The Unitive Equation predicts that it will utilize any spare gravimetric potential it has to interact with the gravity field of the dense mass zone, provided, that the dense zone has some spare gravimetric potential to utilize; right? Not quite!

3.3.3. REGRESSION:

3.3.3.1 First Premise:

Even though objects have a finite amount of inertia-entropy to utilize for gravimetric interactions, they also simultaneously always have to utilize their gravimetric potentials to 100% capacity (both internally and externally). Newtonian physics would collapse if that were not so; however, this is a quantitative statement only. Which exterior mass zones an object's gravimetric potential will interact with is determined by: its total external free energy available, its macro environment's free energy available, and its geometric arrangement and motion relative to surrounding objects. An object can undergo an internal *elastic* transformation in which it can either gain or lose mass by becoming more or less inertial respectively; this transformation, however, is imposed on the object by its macro environment. Furthermore, an object can undergo a similar internal *elastic* change in its states of entropy, i.e., its kinetic energy can be transformed into flux energy and vis-a-vis. (If there is to be an interaction of particle to media, both inertia-entropy ratios have to be compatible. *Elastic* refers to the process wherein the sum of the object's (inertia + entropy) = a constant before and after, and therein the Unitive Equation is born.)

First Premise:

> *An "object" simultaneously has internal inertial and entropic components which can be stimulated or transmuted by natural-outside-energy gradients; thus, these internal inertial-entropic parameters will sustain the object's existence while simultaneously responding to its environment.* This is where the author and classical physicists disagree. I contend that because substance (gravity flux both internally and externally) always occupies space (has form); substance has destiny, or substance need never be treated as a *scalar* function, but can always be treated as a substantive vector; a vector of substance.

3.3.3.2 Second Premise:

> *An object has the internal ability to coalesce free energy when its environment permits such energy exchanges, and vis-a-vis an object is subject to giving up energy when its environment's need to utilize it is greater for existence than the object's need to utilize it for existence.* This may appear to be a redefinition of the Second Law of Thermodynamics, but it is not. Rather it is a recognition that coalescence and dissipation of energy to-and-from an object, from-and-to its environment is natural for existence and coexistence, and that an object by itself is in a constant state of subjugation to its macro environment.

Example: If the object in question is a Bluebill duckling, its potential to coalesce mass is positive, and the duckling will order mass from its environment to become an adult Bluebill. The duckling's potential to become more inertial, *inertial potential*, is positive.

Motion as phenomena involves the interaction of both of these premises and may be as simple as a ball falling, complex as a photon traveling at c, or you taking a break by heading for the frig for a snack. All objects get self perpetuation, *life during their existences*, from the interaction of these two premises.

In using the Unitive Equation, it is not enough to know the ratio of internal inertia-entropy, one also needs to know if the object is in a state of growth or decay to determine the object's destiny. On these two premises rest all objects' existence and coexistence.

3.3.3.3 Redefinition of Inertia:

In the past, the definition of inertia meant that a mass at rest or a mass in motion has a stability about it due to its lack of motion or directional motion respectively. The author finds this definition too confusing because each state of stability means the opposite of the other when dealing with an object.

The author's definition of terms are as follows:

Inertia is the property of objects to resist acceleration or motion.

Entropy is the property of objects to resist deceleration or change of direction of motion. (Actually Newton's definition of inertia mirrors the Unitive Equation because an object always unitively has both types of his inertia.)

Why the redefinition?

The author was taught that inertia was a measure of order and entropy was a measure of disorder, and that is wrong! Even though by classical definition this is so, by unitive thought, disorder is by definition dependent on motion. Because disorder comes about by motion, and inertia also comes about by motion, I have concluded that entropy and the inertia defined in the second portion of Newton's statement are one in the same (kinetic energy).

An increase in inertia or entropy can bring about both stability and instability. The tenacious rigidity of our universe is born out of these complex gravity flux interactions, forming the fabric of the universe. The Bible calls it a firmament.

EXAMPLES:
Examples of when inertia brings about order and when entropy brings about disorder are historically evident, but two examples where their inverses are apparent follow respectively:

1. Our sun is too inertial (ordered), and it is stabilizing by decaying (disordering) its mass; and

2. A spinning gyro becomes more stable as its angular velocity increases.

NOTE: That the perspective of both is the same, we give the object a sense of destiny and by exterior (transcended) examination, we conclude its destiny by its use of its own inertia-entropy in relation to its environment. We are not concerned at this moment why the sun became too fat, or how tired the donkey is because he made the gyro go faster! Only how the object behaves to its environment.

3.3.3.4 Reflection Problems:

Example:

A ball reflecting off a wall:

Now using the same perspective, let us examine a ball reflecting off a wall. As the ball strikes the wall, all of its kinetic energy (that portion of its entropy we are observing) comes to rest. The ball's kinetic energy is inertialized by its environment. Because the ball's state of inertia is too large, it will use the energy stored in its strained bonds to resume its former shape, as both it and its kinetic energy are reversed.

3.3.3.5 Gravity:

GRAVITY PROBLEMS:

To a more complex problem regarding this phenomenon:

When I took physics in college, I was taught that a man weighs less at the equator, than at the north pole because the distance from the man to the earth's center is less at the pole. I believe that to be an over simplification of what really takes place. Resnick - Halliday, Physics I section 16-5, page 395–397. On page 395, Table 16–2 lists the acceleration of gravity, *g*, for *0°* latitude at *9.78039 M / sec²* and *90°* latitude at

22

$9.83217 \, M/sec^2$. The text goes on to say that tangential acceleration accounts for $0.0336 \, M/sec^2$ difference, leaving $(9.83217 - 0.0336 - 9.78039) = 0.01818 \, M/sec^2$ difference due to radii differences; but, $dg = Gm(1/R^2_{pole} - 1/R^2_{equator})$ = $0.06478 \, M/sec^2$, or over three times the value needed.

If we look at the problem in a different light – that an object can gain and lose mass depending on, and depicted by, its inertia-entropy ratio – we can calculate that a one kilogram object at the pole would gain kinetic energy as it is moved toward the equator. The gain in kinetic energy would be:

$K_e = 1/2 \, I\omega^2$,

$I = mR_o^2$.

R_o = distance from the object to the earth's axis of rotation, approximately = $R_{lat.}(cos\theta)$

$K_e = 1kg\omega^2r^2 / 2 = 1kg \, (2\pi/T)^2 \, (R_o^2) / 2$,

$K_e = 1kg \, [(2\pi)^2 / (86,400 \, sec)^2] \, (6.378 \times 10^6 \, M)^2 / 2$,

$K_e = 107,600 \, kgM^2/sec^2$ at the equator.

We can equate this loss of inertia or loss of mass as:

dm (change in mass) = $(dK_e) / c^2$,

$dm = (107,600 \, kgM^2) / (3.0 \times 10^8 M/sec)^2$,

$dm = 1.2 \times 10^{-12} kg$.

This rationale leads to the concept that both the mass of the one kilogram object, and proportionately the mass of the earth are less at the equator. The Unitive Equation adds another dimension to this problem based on the concept that *inertia = (mass + lack-of-motion-of-mass)*, and *entropy = (free-gravity-flux + kinetic energy)*. This new dimension is thought to be important when studying life systems. Suppose this one kilogram mass was a Bluebill duck in late September

23

near the north pole. Having spent the summer subjugating energy from various food organisms (which in turn subjugated solar energy), the duck then utilizes this mass to gain kinetic energy to fly towards the receding sun [it knows (?) what side of the world its bread is buttered on]. [Note that if this problem is viewed from a point exterior to our solar system (transcended) there is no change in mass (mass from sun, to earth, to organisms, to duck and then to duck's environment; however, internally this process represents dissipation of mass, and the sun's mass ends up being less inertial because now it is revolving about the sun etc.)]. We need the mathematics to allow our duck to preserve his existence within the freedoms afforded to it by its environment, or *(1) chubby duck = (1) leaner duck + (1) leaner duck's linear kinetic energy + (1) leaner duck's rotational kinetic energy.* Our Bluebill did not have to ask the sun, the earth or the organisms to fly south, but our duck cannot grow to *100 kg.*, or fly at *500 km.p.h.* either.

Figure 3-2:

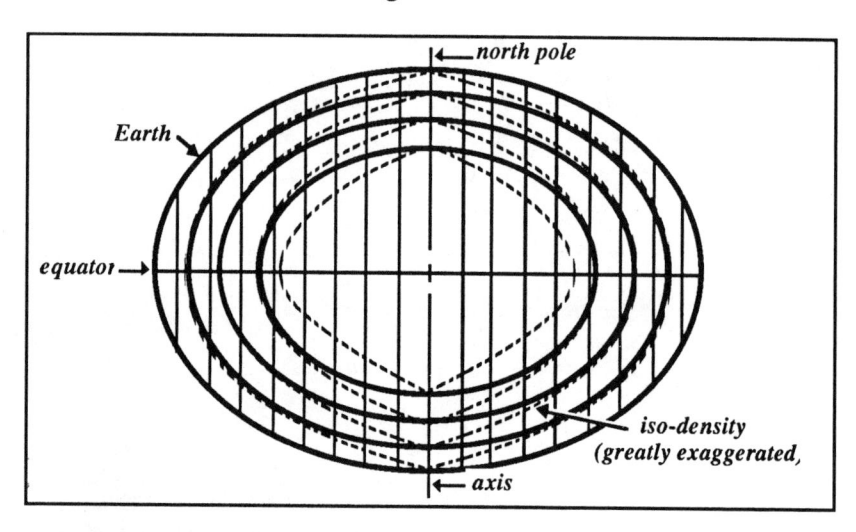

Figure 3-2: Mass distribution of Earth based on loss-of-mass due to kinetic energy complexed with a quasi-linear increase in density toward the Earth's center.

This rationale depicts that the density of mass distribution within the earth is a function of two variables, i.e., the density increases towards its center by a quasi-linear function, and decreases in density as a cylindrical function around its axis with increase in radius. Refer to *Figure 3-2*.

Newton's equation may be:

$$F_g = AG\,(m_e m_o) / R_o^2 - G(m_i - dm_i)(m_o - dm_o) / R_i^2 - (m_o - dm_o)\,a_r.$$

Where:

$A \leq 1$	decreases with decrease in latitude.
$R_o =$	Earth's Radius @ object.
$R_i =$	Not an exact figure because it depends on inertial (polar equivalent) mass of object. Conceptually it is the effective mean-radius of all the earth's mass entering into an isostic interchange. Most likely $(m_i - dm_i) = (m_o - dm_o)$ as indicated by the following problem with earth tides.
$G =$	Universal gravitational constant.
$m_e =$	The earth's mass.
$m_i =$	The earth's inertial (polar equivalent) mass entering into the isostic exchange.
$dm_i =$	Change of m_i due to gain in kinetic energy.
$m_o =$	The object's inertial (polar equivalent) mass.
$dm_o =$	Change in m_o due to gain in kinetic energy.

Note: That Einstein's equation $E = mc^2$ was used for this inertial decreasement, the Unitive Equation works out as well,

$$E_t = mc^2 \cos^2\theta + E_f \sin^2\theta,$$

because $E_f \sin^2\theta$ is not a mass function.

We are only interested in loss of the mass function, and in the true light of the equation $E_f \sin^2\theta$ increases as there is a decrease in mass function due to loss of inertia for both the one kilogram object and the earth's equatorial mass.

Why complicate Newton's equation?

The rationale of a man weighing more at the pole of the earth because he is geometrically closer to the earth's center is not valid; however, this rational would be valid if one said because at the pole he is closer to the dense center of the earth, or even that the bulk density of the earth at the pole increases more rapidly between himself and the earth's center. While it may be true that the various density gradients within the earth may mirror Newton's gravitation laws, the geometry of this problem would seem to cause an inverse relationship – as the pole radius gets smaller the total amount of mass pulling a polar object directly down decreases, as the total mass pulling the object radially out toward the equator increases. If we theoretically flattened the earth to a polar radius of one kilometer and place a man on the pole, most of the mass of the earth would be pulling him radially outward. Would the man gain mass?

My answer to that question is:

He better, or he will cease to exist as an object. Will he cease to exist as a human being? Yes! Will he cease to have life? Later in this paper I will "approach" quantifying life with this Unitive Equation. After all, if this equation has any starch in it, we should be able to utilize it for quantifying life too; perhaps other physicists have already done this?

The centers of all mass aggregates have their own black holes, so to speak — that region where all net-cancellations of gravity energy of the aggregate's objects are centered around. These net-cancellations of energy vectors give rise to inertia. This usually is expressed as the object's mass; however, the quantitative mass density of this center is also related to how much gravity flux is being used by each object of the aggregate to support the aggregate's being or existence within its macro environment. Our galaxy is a unitive system to the universe with its location, its motion, its spatial volume, its inertia-entropy ratio, and its internal behavior all dependent on elastic responses to its environment. The same is true with our solar system's relationship to our galaxy, and so on, down to the smallest of particles. How does this affect our theoretical man standing at the center of our flattened earth? Much of his entropy will be changed into inertia; thus, he will gain mass or become more dense, and as a species of matter, his human life will not exist at this density. As an object, will he cease to have life? All objects have life and death by their destinies. I concluded that *life* is *sustenance of an object's physical / metaphysical identity* with plant and animal life being separate harmonic scales within the objects of the universe; therefore, if our man cannot remain an object while he transfers his entropy into inertia, he will die, or his substance will be coalesced by other objects about him. I will address spiritual existence later.

[In editing my work, I thought I should do more than leave you with the above equation which may not be suitable for pragmatic use. I ran several comparisons between *g* (*g* is what the reader knows as the acceleration of gravity at the earth's surface) which may be found in **Table 3–1**. In column *1* the latitudes are listed; in column *2* the radii of the earth by latitude are listed which were approximated by solving the following equations,

$R_{lat.} = (x^2 + y^2)^{1/2}$, where;

$x^2 / R_e^2 + y^2 / R_p^2 = 1$, and $\tan^2(lat.) = (y/x)^2$;

given $R_e = 6,378,390\ M$, and $R_p = 6,356,910\ M$.

In column 3, the values for *g* from Physics, by Resnick and Halliday, Part 1, Table 16-2, p 395, John Wiley & Sons, Inc., Third Printing 1967 are listed; in column 4 the values for *g* calculated by the Newtonian equation $(g = Gm / R^2_{lat.} - ar)$ where $ar = (2\pi / T)^2 R_{lat.} \cos^2(lat.)$ are listed; in column 5, the values for *g* calculated by the "International Gravity Formula" from Introduction to Geophysical Prospecting, Dobrin second edition, McGraw Hill Book Co., 1960, page

234 are listed. In this equation I altered *g* at *0°* latitude from *9.78049 M/sec²* to *9.78039 M/sec²* such that:

g = 9.78039 M/sec² [1 + 0.0052884 sin² (lat.) − 0.0000059sin² (2lat.)]

to match *g* in column 3 at *0°* latitude. In column 6 the Unitive results are listed as explained; in column 7 the percent variance are listed between column 3 and column 6, or [(column 6)/(column 3) - 1]x100% to five significant figures.

The Unitive Approach:

I do not know how the figures in column 3 (Resnick-Halliday) were obtained, but because they are listed to six significant figures, I used them as the standard. I believe *g* and therein *G* are differentials unto themselves, or *g* at any latitude can be expressed as: $g_{lat.} = gd(g)$. If we allow *G* to vary with latitude, we can use the figures in column 3 as empirical data. From the Newtonian equation let *g = Gm/R²;* where *G* is a variable, and *d(g) = d(g)cos²(lat.) + d(g)sin²(lat.),* or a closed system. *g* will vary with radius and with *G*. If we calculate *G* at the equator, this will be the *G* with the highest amount of entropy in our inertia-entropy ratio. As we move toward the pole we will gain inertia which is represented by:

d(g)cos²(lat.)= d(g) − d(g)sin²(lat.), or

d(g)cos²(lat.) = d(g)[1 − sin²(lat.)].

In a closed system of dot products where,

E_t = Δ*(g)* = *(total variance),*

cos²(lat.) = d(g), and

d(g) = [1 − Asin²(lat.)],

where *A* is an environmental "constant" of how the *cos²(lat.)* will vary with the *sin²(lat.)*. *A* includes such factors as density, shape, rotational velocity, orbital velocity and other cosmic inertia-entropy variables.

Note: *A* at times may be something other than a numerical constant; it may be a geometric in and of itself. In a closed system unity *(1)* stands alone, and the $sin^2\theta$ function must vary as a dependent variable to the $cos^2\theta$ function. Multiplying this equation by the Newtonian equation with a variable *G* follows:

$$g_{lat.} = gd(g) = (Gm/R^2_{lat.})[1 - A sin^2(lat.)].$$

If we allow $B = (Gm/R^2_{lat.})xA$, the following equation results:

$$g_{lat.} = Gm/R^2_{lat.} - B sin^2(lat.).$$

We first calculate *G* at the equator using the value in column 3 at $0°$ latitude as follows:

$$9.78039 \ M/sec^2 = G_e \ (5.983x10^{24} \ kg) / (6,378,390 \ M)^2 - 0; \ G_e = 6.6506 \ x \ 10^{-11} \ nt\text{-}M^2 / kg^2.$$

Using this value, we calculate *B* at the pole using *g* from column 3 at $90°$ latitude, and the polar radius from column 2:

$$9.83217 \ M/sec^2 = \frac{(6.6506 \ x \ 10^{-11} \ nt-M^2/kg^2)m_e}{(6,356,910M)^2} - B(1),$$

$B = 0.01443M / sec^2$. Substituting, the resulting equation is:

$$g_{lat.} = \frac{(6.6506 \ x \ 10^{-11} \ nt\text{-}M^2/kg^2)m_e}{(Radius \ of \ Latitude)^2} - \frac{0.01443M sin^2(lat.)}{sec^2}.$$

Using *R* from column 2 the reader should verify the results listed in column 6 from $10°$ to $80°$ latitude. If column 3 represented empirical data, this equation would indeed match the "data" quite well. This equation does not take into account local variations in *g* due to mountain ranges, ocean basins and isostic variations in the earth's crust.

Note: *B* as calculated above *numerically* equals Planck's second radiation constant, see page 50 for further discussion.

TABLE 3-1:

1	2	3	4	5	6	7
Lat.	Earth Radius	g Tab.16-2	g Gm/R^2-ar	g I.G.F.	g Unt.Eq	(6/3-1) x100
0°	6,378,390	9.78039	9.77522	9.78039	9.78039	0.000%
10°	6,377,739	9.78195	9.77824	9.78194	9.78198	0.000%
20°	6,375,866	9.78641	9.78695	9.78642	9.78648	0.001%
30°	6,373,000	9.79329	9.80028	9.79328	9.79336	0.001%
40°	6,369,488	9.80171	9.81662	9.80170	9.80181	0.001%
50°	6,365,759	9.81071	9.83401	9.81069	9.81079	0.001%
60°	6,362,259	9.81918	9.85034	9.81914	9.81924	0.001%
70°	6,359,411	9.82608	9.86365	9.82604	9.82614	0.001%
80°	6,357,555	9.83059	9.87234	9.83055	9.83064	0.000%
90°	6,356,910	9.83217	9.87535	9.83211	9.83217	0.000%]

Table 3–1: Comparison of "g" at the Earth's Surface

So when I look at these gravity problems of men and ducks having the ability to alter their inertia-entropy ratio to maintain their identity, I conclude that their mass actually does change; furthermore, this phenomenon is another expression of gravity (the object) having the metaphysical ability to interact with surrounding flux vectors in its environment. Herein lies a dilemma. Do we quantify objects by existence, or by inertia-entropy? I prefer to quantify objects by existence which have varying amounts of inertia-entropy. The reason is that as an observer, objects and their various behaviors are perceivable, but I may never be able to view an object's true inertia or entropy. Why? Objects are always in a state of absolute motion; and therefore, objects have quantative amounts of absolute inertia-entropy. I believe predictability will come by coupling an object's existence with its inertia-entropy. *Existence has been overlooked as not having physical consequence to physical law.* Your toe still exists, even though the molecules (mass) which once hurt so much when you stubbed it have since left you.

3.3.3.6 Tides

EARTH TIDES

Earth tides would also appear to be caused by an isostic behavior of the earth's mass to the moon. An interesting question is: why are tides a meridial phenomena? Refer to **Figure 3-3**.

FIGURE 3-3:

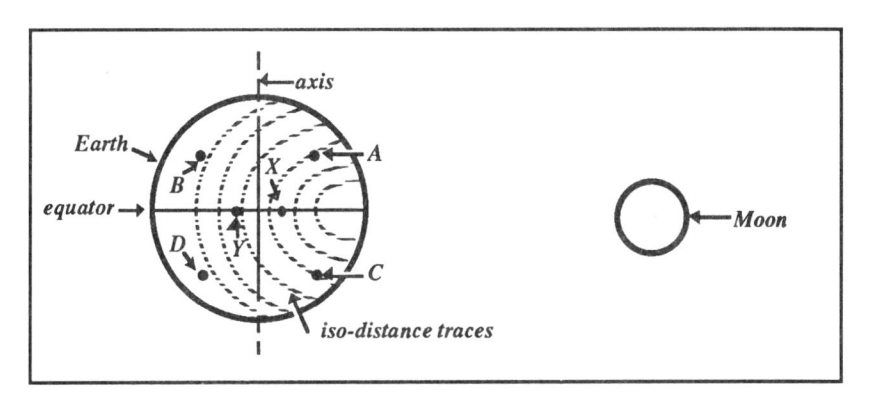

Figure 3-3: Showing iso-distance traces from the Moon to the Earth's surface.

The moon and earth must utilize their total gravity flux to 100% while interacting. Note that these iso-distant traces from the moon to earth are elliptic with their long axis parallel to the equator. Note also that the earth's mass near the equator suffers a state of more weightlessness due to its increase in tangential acceleration. Now if mass is a scalar function, earth tides should be an equatorial phenomena, rather than a meridial phenomena base on this geometry and weightlessness.

Let's examine this problem on the premise that the earth is an object which has the freedom to respond to variations in its environment by varying its inertia-entropy ratio. In other words, we give the earth a sense of being. Also, because of the constancy of the earth - moon system to the sun and therein our galaxy, let's ignore that gravimetric flux utilized for their existence within this larger system. We are only interested in the utilization of that part of earth's and moon's gravimetric flux which brings about a unitive or closed system.

In reference to **Figure 3–3**, the moon will pull harder on points **A** and **C,** than on points **B** and **D** which are symmetrically opposite, or *A* and *C* will have a higher state of inertia placed on them by the moon utilizing its gravimetric potential as an inverse-function-of-the-radius-squared from the moon to these points. Also, the center of mass for this earth - moon system lies somewhere between the earth's center and its equator. The location of this *black hole* is both due to the amount of matter involved and also due to an increase in inertia due to the net-cancellation of the earth's and moon's gravity energies. Let's assume it is located at position *X* in **Figure 3–3**.

The above (per this Unitive Theory of net-cancellations) herein postulates why a higher state of inertia exists on the side of the earth nearest to the moon. So why are tides meridial?

This state of higher inertia is a forced state onto the earth by the moon. The earth could become more inertial by moving its water to a higher latitude where its water has a lower state of entropy or kinetic energy; however, this makes no sense because the earth's waters would be traveling farther from the moon which caused its higher state of inertia. Also, the water will not be pulled down latitude because that puts this water in a higher state of entropy which is opposite to the inertialization process impinged upon the earth. The only option left is the orthogonal one – the waters of the same latitude will respond to this inertialization process. An alternate perspective of our earth's response might be that the mass or waters of the earth with the same inertia-entropy ratio (like vector systems) will interact with each other to bring about this inertialization process.

Why is there a tide symmetrically opposite to this meridial tide? Because points *B* and *D* have to utilize their gravimetric flux to 100%, there is a symmetrically opposite inertialization of the earth. In other words, when points *A* and *C* use their gravimetric flux to interact with the moon, they do this by interacting less with points *B* and *D*; thus, *B* and *D* have residual gravimetric flux to interact with like-neighboring-vectors causing a higher state of inertia symmetrically opposite of points *A* and *C*. The earth due to its shape and rotation must distribute its mass function, so that an opposite *black hole* at *Y* will appear.

THE MOHO

The author believes that Mohorovicic seismic discontinuity underneath the earth's crust is a function of the earth's crust responding isostically to the moon while the interior of the earth tries to rotate away from its crust. (Similarly, certain continental drifts may be explained this way if it can be shown that the remnant magnetism in a West plate depicts a faster spreading velocity to the west than an East plate's velocity to the east.) The location of the Moho appears to be a function of plasticity due to increasing density (in the form of heat and pressure) from the earth's surface, and also to the amount of mass needed by the moon for this isostic gravity exchange. It would appear more complex than to just say the Moho occurs at the shallowest zone of plasticity; it may occur deeper if there is insufficient mass for this isostic exchange. If we assume the Moho to be 50 km deep and that the density of the earth's crust including oceans to be *2.80 gr./cm³*. (Resnick - Halliday, Physics I, Table 17-1, page 425.)

The mass of this outer shell of the earth is:

$M_s = 4/3 \pi d [R_e^3 - (R_e - 50km)^3]$,

$M_s = 4/3 \pi (2.80 gr/cm^3 / 1000 gr/kg) [(6.368 \times 10^8)^3 - (6.318 \times 10^8)^3]$,

Mass of earth's crust above the Moho = 7.08 × 10²² kg (atmosphere not included).

Mass of moon = .01228 earth masses (Resnick - Halliday page 2, appendix C, Part I).

Mass of moon = .01228 × 5.98 × 10²⁴ kg,

Mass of moon = 7.35 × 10²² kg — Pretty Close!

The author realizes that *50 km* is not the average depth of the *Moho – 33 km* is probably closer, but it does extend to *50*

33

km in some areas, and would seem to get much shallower where the fluidity of the earth's mass due to tidal inertialization deforms (either temporarily or permanently) the earth's surface. Also, the solid portion of the earth's inner core is fairly equivalent in mass:

Inner core mass $= 4/3\pi d\, (1.250 \times 10^8)^3 / 1000 gr / kg$,

Inner core mass $= 7.77 \times 10^{22}\, kg$,

when $d = 9.5 gr / cm^3$.(Resnick - Halliday, Physics I, Table 17-1, page 425.)

What is the meaning of this?

Perhaps gravity has surface charge characteristics like static electricity? (Charge is not a fundamental unit in this Unitive Theory - electromotive force is a potential of mass for motion, or potential for occupation of a spatial region and can be described as a gravity interaction, not an electrical one.) Is the moon *negative* and the meridial tide near the moon *positive*? No! Is the tide symmetrically opposite *negative*? No! We can utilize gravity phenomena to explain "*positive*" and "*negative*" distributions.

Also, if the Moho is a function of the earth responding isostically to the moon, we have to come to the conclusion that gravity interactions have a bulk shear modulus through space. This would appear evident by the rigidity of our universe. Also, by the fact that light has a top-end-velocity — no physicist has ever well explained to me, why light, needing no media to travel - can only travel a *finite* rate of speed! Why? This finite velocity would appear to have most of its existence with some type of shear interactions orthogonal to the line of propagation because all energy levels of photons reach this *finite* velocity. If this *finite* velocity were caused by resistance to spatial penetrations by photons, the large quanta with more momentum would win out, but when white light is refracted by a prism, the opposite phenomenon happens - i.e., the higher the frequency, the greater the bending by the prism. Past equations have validated these higher indices of refraction based on slower velocities of the higher frequency photons while in the prism media. Though valid, the logic of these equations seems a redundancy unto itself. A better question to understand the nature of light might be - why do photons suffer velocity changes proportional to their frequencies while being refracted by a prism? (This will be addressed later.)

When we look out into space, we have a tendency to base most of the motion seen there on gravity interactions. Our earth-moon system is in a state of gravity oscillation while simultaneously our earth-moon system is in gravitational orbit about the sun; however, when we try to explain the unseen forces that bind matter together - we have coulomb forces to explain why the positive protons of our atom's nuclei do not repel and fly apart; we have electrostatic forces that bind our electrons to our protons; and we have Van der Waal's forces to bind our atoms and molecules each-to-each. The author has come to the conclusion that coulomb forces, electrostatic forces (including the inertial potential of magnetism), and Van der Waal's forces are all expressions of gravity interactions. (Charge with its electro-magnetic characteristics will be discussed as a gravity phenomenon later.)

As to the *coincidence* of an equivalent amount of earth mass associated with the moon above the Moho, and at the earth's inner core, does *surface charge* phenomena of gravity cause nucleation at the center of the earth? Radical; but keep in mind that the dilation I spoke of earlier describes a **black hole** on either side of the center of the earth at *X* & *Y* in **Figure 3-3**. Perhaps this dilation exists as a spherical nucleus and that our concept of a fluid/solid interface at the inner core's outer boundary is in part an interference layer. This later thought came to me some 3 months after the concept of the Moho being caused by the earth's and moon's gravity having *surface charge* characteristics. So I did the calculation, and the inner core's mass seems within 6% of equivalence to the moon's mass. Computer modeling of this problem may prove it wrong or give us more insight into this concept. Nucleation is at the hub of many physical processes. This may be a form of it. In the abstract sense, a nuclear equivalent (of the exterior gravity used by the earth to interact with the moon) condenses much like a **black hole**, or the **packing fraction** of the earth's nucleus increases as a function of the total gravity impinged (net-cancelled) on its centroid.

3.3.3.7 Gravity Energy?

We have always related gravity to the mass function of an object, but is gravity, free energy? It would seem it would have to be by its nature to occupy space. The Unitive Equation allows for such a co-existence due to its simultaneous pairing of mass-energy vectors (a format of substantive vectors occupying space). Can we find an equation that will give gravity an identity and therein show that mass is inertialized gravity, and gravity is *free* energy occupying space and available for work expressions? *Free* meaning tied-up or existing exterior to the object.

Discussion:

Energy comes to us by a historical definition as the capability to do work, or *E = force* x *distance.*

Mass comes to us by more than definition, objects are all about us, and in that light Newton gave objects relative mass by his equation *F = mg*; thus, by the variations in *F*, we give these objects their respective masses. The problem with this equation is that it is as relativistic as Einstein's – *E = mc²* equation. It simply is not an inertial equation. To prove this, let's theoretically stand on the sun and watch a ball "fall" back to the earth. We can see that the force of gravity pulls the ball and earth together and that this force is in operation over an orbital path length *x* of both the ball and the earth as these two objects orbit around the sun. Also this net-cancellation of forces occupies a space exterior to both objects (gives form to space); but, like a *black hole* its centroid lies specifically between these two objects; thus, from this point of observation, gravity is not a force, but a force applied over a relative distance *x*, or

Eq. 3.3.3.1 $gravity = F(x) = Gm_e m_o(x) / R^2$.

The sun must do work to move this net-cancellation of forces through its orbit. Note the kinetic energy of the ball is not "created" (it was in the system all the time and is not available for us to do work on the sun – the same is true when our Bluebill flies south for the winter). The author postulates that *force* is an *internal* expression of free energy at work, or that you can never have a *force* without free energy being present. Much like momentum is the first derivative of kinetic energy, so too the *force* Newton gave us in his gravity equation is the first derivative of gravity energy.

Suppose while we are standing on the sun we examine the internal substance of the ball. If we examine the *relative motion* therein, by the same rationale, we would conclude the gravity therein can be expressed as:

Eq. 3.3.3.2   Gravity internal

$$= \sum_{1}^{n} F(g)_n \, (x)_n = \sum_{1}^{n} m_n \, g_n \, (x)_n \ .$$

Where: n =   The number of internal objects in motion,

x_n =   Their "relative" respective distances traveled, and

g_n =   Their net-cancellations of internal gravity forces exterior to these internal particles but also internal to the object.

The term *relative motion* refers to each object's motion relative to its paired gravity centroid; thus, we don't have to complicate this equation by the fact that the ball is actually falling, rotating with the earth, orbiting about the sun etc..., because we have transplanted Newton's physics to each of these events. However, in absolute terms, we do need to recognize all motions of these internal objects. In physical terms, Newton (ignorantly perhaps) recognized the sovereignty of these internal net-cancellations by calling an object's *lack - of - freedom - to - use - its - interior - gravity - for - exterior - net - cancellations — its mass!*

What this really means is the *inertial mass* of the object is not *identical* to the *gravity mass* of the object. For all practical purposes at natural or equilibrium conditions, these two masses will appear equivalent in Newton's reference frame, or gravity (g_n) = mass (g_n); however, in Einstein's reference frame they diverge which is evident by Einstein's equation, $E = mc^2$, being twice as large as Newton's kinetic

37

energy equation, $K_e = 1/2mc^2$. Einstein's equation correctly states that the amount of energy given to a photon as it leaves a decaying object (our sun) will be twice the amount experienced by a receiving object (our chubby duck). The other 50% ends up as *active energy* or *gravity energy*. With this rationale, there is no room for charge or magnetism as *fundamental substances* because neutralized charges end up as stationary energy (mass) which must have a gravity field, and if gravity energy can net-cancel to form mass – can we form mass from two different fundamental substances? Also, objects of mass can exist without exhibiting exterior charge or magnetism–but always have a gravity field for interaction. I believe that rather than chase this riddle around from charge, to energy, to mass, to gravity, to magnetism, to charge, etc..., recognize the universe is made up of one substance, *gravity energy*, with the properties that it can exist in several types of manifestations.

Gravity = The ethereal mass of the universe,
Mass = Inertialized gravity,
Electromotive force = Potential for mass to be exteriorly entropic,
Magnetic force = Potential for mass to be exteriorly inertial.

This rationale predicts – the more gravity required externally by the object's environment, the less mass an object will have internally. Conservation of this mass loss will be perceived as lack–of–motion or inertia. Large heavenly bodies require rapidity of motion from smaller bodies, and vis–a–vis smaller bodies require the exterior consumption of gravity from large bodies which inertializes these larger bodies. If this process of aggregation continues by addition of more and more mass around the centroid a *black hole* is formed. Depending on surrounding environmental conditions a *black hole* may be very small, or a large cosmic event. Mass is equivocated to inertia because mass and inertia represent the *ability* of gravity to occupy an interior spatial zone. To a classical physicist mass is scalar and not synonymous to inertia (which is stationary entropy), but that only appears true in the reference frame in which you apply Newton's sovereignty.

Active gravity and kinetic energy represent the *inability* of gravity to occupy an interior spatial zone. These are responsible for the mass diffusion characteristic of the universe.

3.3.3.8. The Theory of Relativity vs Wave Mechanics.

Yesterday and today while my work was in editing, I relaxed and read Milton Dank's, *Albert Einstein*, recommended to me by a friend. Today I discovered what Einstein's "unified field theory" proposed (I agree that "charge-magnetism" are gravity manifestation, – I have for a number of years). Today I also found that Schrödinger believed in net-cancellations, as I do, never having read his works I don't know if his logic is the same as mine. [Schrödinger was mentioned in my 1978 works because my physics text (Resnick & Halliday, p.1204–1205) indicated he predicted reflection would take place at a node.] Today, I had to retrieve my work to rewrite this section, so that it *unites* the Theory of Relativity and Wave Mechanics into one. Because like an Einstein, I stood on their graves years ago and "decreed 'go to hell'", my work never took on their prejudices; and therefore, I didn't write this work to iron out the wrinkles between "Jewish" and "Prussian" physics. I honestly believe that had I pursued physics as a profession, I would never have written this work – I would have taken sides.

When we look inside the ball from our sun, we don't see the invisible centripetal energy keeping these cyclical resonant oscillations stable. Because I eliminated charge as a fundamental unit, I had to replace these quasi-stationary energy couples with gravity. In the same manner we can't see the *active energy* holding us to the earth or holding our earth in orbit, but at each higher level this becomes mass. This continual process gives all the smaller particles their spatial (higher order) structure, and gravity ends up occupying space either internally or externally.

So it is, we have to couple these two corporeal entities, *gravity energy* and *mass* with two meta–corporeal (corporeal – incorporeal) entities, *distance* and *time*. We can sense what distance and time are, but we cannot put a velocity vector in our pocket.

Einstein combined these "principles of equivalence" parameters to state that in a relative reference frame gravity and acceleration effects are equal. Because gravity was not considered a substantive–scalar–equivalent of mass, he equated gravity to acceleration. Einstein's "thought experiment" of a box containing a woman and a ball accelerating through space is really a stimulated pre-adventure; the equivalence actually lies between the kinetic energy imparted to the box and contents, and their stationary inertia. (I don't believe in curved space, rather I believe in exterior gravity fields.) Recognize however, in identifying this inertia equivalence, Einstein came up with the correct answer.

The Unitive Theory combines these parameters to state: in a closed system gravity and mass are transmutable, as well as time and space are transmutable; however, one should not expect a transmutation between the corporeal and the meta–corporeal. They are not *like* vectors and will not transmute. Rather, the relation between these two dual function systems describes a third entity–spirit, or life properties. In my 1978 work I approached this third entity as *inertial potential* and *entropic potential*, or life within (i) and life without (e). This observable dual parameter provides mass and gravity with the essential metaphysical characteristics of interaction within their own dual parametric self and depict the inter–dependent relationship between the corporeal and meta–corporeal. This spiritual entity is totally an *incorporeal* harmonic to nature, with *inertial* and *entropic potentials* depicting this spirit's or existence's behavior. This life entity may be active in you as you read these words – you may feel that I am toying with your dignity! Because your existence vector has such a long cycle, your toe can suffer many mass changes (much like our sun pulling in cosmic debris while simultaneously radiating out solar energy).

From our transcended frame on the sun, the internal net–cancellations within the ball occupy space and are not available for mechanical work on the sun. They are also only slightly available for mechanical work to someone on the earth. The concept being: the ball's internal gravity is even further removed from us for mechanical work because it operates in a freedom afforded to it by the collectivism of its immediate environmental constituents. Similarly, our Bluebill did not need to ask anyone to fly south. This lack of freedom to the ball's constituents and freedom to our duck equates to Wave

Mechanics "probability" – "uncertainty" of locating these respective objects.

Example: From our transcended frame we examine all the planets for our Bluebill. If we use a resolution power where we can only see the northern hemisphere of the earth, and 100% of the time that we locate our duck, he is on the earth;

> The duck's freedom to live elsewhere (another planet) is; $1 - 1.0 = 0$.
>
> However, the duck is only locatable 98 days of every 100 days that we search the northern hemisphere;
>
> Thus the duck's freedom to live elsewhere is: $1 - .98 = .02$ or 2% etc...

The only entity that can express why there is a difference in the above probabilities is life or spirit, no corporeal or meta-corporeal entity can explain why this one kilogram object does not always have to exist in the northern hemisphere. The reader may be reluctant to recognize life within atomic or subatomic resonances, but that is exactly what every physicist does when he/she recognizes the *probability–uncertainty* of these resonances.

Even though this seems that I am pitching my tent on the side of the Wave Mechanics promoters; I have to pitch my tent on the side of that "Old Jew" and "Old Gentleman" because you will never convince me that on those two days / 100 days that we can't find the duck in the northern hemisphere – the duck doesn't exist or, the duck can't see our transcended frame, the sun. Unless this inverse probability realistically exists, we need to express our 2% ignorance in some other way. Life is a reasonable way; after all life explains all ignorance! Note, if we decreased our resolution power and examine the whole earth at once we may never see our duck; thus our *uncertainty* (lack of faith) has altered the meaning of existence.

3.3.3.9 Uniting Historical Physics

One of the inherent problems with the historical definition of energy, as defined by classical physicists, $E = f(x)$, is that energy is only defined for the system which utilizes it for *mechanical work* purposes; all other forms are ignored. In abstract terms this puts our corporeal vectors' existence at the mercy of our meta–corporeal vectors (time and space) which is wrong because transmutation will not take place between these systems; they are simultaneously inter–dependent and intra–independent systems. The law of conservation of substance excludes substance being derived from meta–corporeal entities; thus, the definition of $E = f(x)$ is poor because energy's existence is not derived from motion; rather, *force* is dependent on free energy as its first derivative.

The historical interpretation of energy comes to us by Newton's concepts of action and reaction; therefore, if you do not see an action; you do not expect a reaction or force present. Thus when Newton formulated his equation, he did not account for the net–cancellation of gravity, or gravity energy between the ball and the earth that kept his problem relative, or *Newton's force (internally viewed) = Energy = (F)dot(x) (externally viewed)*. Early physicists believed that the inertial mass and the gravitational mass were synonymous – they appear to be in Newton's relative frame, but not in our transcended frame. The net–cancellation of gravity between the ball and the earth are constant to the sun, and represent mass to the sun. The inertia represented by the ball resting on the earth was equivalent to the inertia represented by the ball when it stopped at the apex of its trajectory. The potential energy (inertia/mass) gained by the ball as it reached its apex was not "created" such that:

$$(m_e + m_b) \rightarrow (m_e + m_b + E_p).$$

Rather, this (inertia/mass) was the result of the earth and ball having the *metaphysical ability* to regulate – stimulated entropy phenomena. This *ability* is our earth's life–within or is an expression of its inertial potential – i.e., the earth's potential to increase its mass by incorporating more mass, (no different than our Bluebill starting out as a duckling). The classical definition of energy has this incorporeal action vector inherent in its meaning; thus, Einstein had to incorporate anthropomorphic characteristics into c to equivocate energy to mass. (Einstein gave c invariable absoluteness.) The Unitive Theory places absoluteness in a unit–life vector:

$$1 = e^2\cos^2\theta + i^2\sin^2\theta,$$

within/between these corporeal and meta–corporeal entities.

In our transcended frame we might approach the Unitive Equation as follows:

Eq. 3.3.3.3 Let $E_t =$ *(inertia) + (entropy)*,

where E_t is a conservative unitive vector. Inertia may be expressed as internal mass function, or $E = mc^2$ (the mass of the ball) plus its inertia due to its lack–of–absolute motion,

inertia = (mc² + mass–at–rest).

Likewise, entropy maybe expressed as the ball's exterior gravity plus its absolute kinetic energy,

entropy = (? + kinetic energy).

Intuitively the *(?)* above might be filled with a *(1/2)mv²* function, but because Newton's kinetic energy is relative to the earth, I am going to change this value to Einstein's equivalent to account for the energy to keep this problem relative. So I postulate:

Eq. 3.3.3.4 $\gamma c^2 \to$ *the classical definition of energy* $\leftarrow mc^2$,

where γ is a substantive vector. Then, *entropy = ($\gamma c^2 + k$)* such that:

Eq. 3.3.3.3g $E_t =$ *(mc² + mar) + ($\gamma c^2 + k$)*.

If we allow E_t to be a unit vector such that *(mc² + mar)* is always perpendicular to *($\gamma c^2 + k$)* then we can activate this equation by a cyclical unit life vector:

$$(1) = e^2\cos^2\theta + i^2\sin^2\theta,$$

where θ is in the direction of E_t, $e^2\cos^2\theta$ is in the direction of inertia, and $i^2\sin^2\theta$ is in the direction of entropy. The dot product yields:

Eq. 3.3.3.5 $E_t = (mc^2 + mar)e^2\cos^2\theta + (\gamma c^2 + k)i^2\sin^2\theta$.

The author postulates that θ will "hover" about **45°** for natural–aggregate systems, such that inertia roughly equals entropy.

Note: The *e²* is the *potential–for–life–without* and vis–a–vis *i²* is the *potential–for–life–within*. The reason why the potential–for–life–without *e²* has been multiplied by the inertia and the potential–for–life–within *i²* has been multiplied by the entropy rest in my former definition of these potentials. Entropic potential is inversely related to entropy and vis–a–vis inertial potential is inversely related to inertia. When our duckling was growing by coalescing exterior mass–energy, its potential to decay was also growing, and the duck reached its maximum mass when its environment would no longer allow the duck any more energy for coalescence. Also, a unitive vector is justified here because the duck's existence is a closed system. Because there are many different rates of coalescence and decay in nature, these rates may not follow the *(1) = Acos²θ + Bsin²θ* format (where *A* and *B* are scalar constants), but it is believed that variable rates can be obtained by summing up an aggregate of closed systems having different phase angles θ. The stability/instability of these aggregate systems rests in the collective stability/instability of life vectors within/without, or in Wave Mechanics; their *probability* to exist. In my 1978 work, I placed *i* and *e* in opposition (the same plane) as inertia and entropy respectively. I have altered this to say *i* and *e* operate orthogonal to inertia and entropy respectively, and now have values between **0** and **1**, instead of **–1** to **1**.

Now if we simultaneously couple inertia with entropy two different equations fall out, but in order to maintain unity, we must allow the phase angle to vary in our life vectors or:

$(1) = e^2\cos^2 B + i^2\sin^2 B$ for our first equation, and

$(1) = e^2\cos^2 C + i^2\sin^2 C$ for the second equation.

Eq. 3.3.3.6 $E_t = (mc^2 e^2\cos^2\theta + \gamma c^2 i^2\sin^2\theta)e^2\cos^2 B +$
$[(mar)e^2\cos^2\theta + (k)i^2\sin^2\theta]i^2\sin^2 B$.

This equation unites the inertial mass function and the active gravity in parametric unity, as well as the absolute inertia and kinetic energy in parametric unity.

Eq. 3.3.3.7 $E_t = [(mc^2e^2\cos^2\theta + (k)i^2\sin^2\theta]e^2\cos^2C + [(mar)e^2\cos^2\theta + \gamma c^2i^2\sin^2\theta] i^2\sin^2C.$

This equation is similar to Eq. 3.3.3.5 in that it couples active substantive vectors with absolute vectors, (inertia–kinetic energy) and (mar–entropy).

The phase angles B and C are inter–dependent and depict a level of freedom encapsulating or one–degree–larger than θ; however, it is postulated that unless B, C and θ tend toward equilibrium, $45°$, transmutation will take place between these levels of freedom. This seems evident by their dot products and cross products. If we only utilize the dot products of Eq. 3.3.3.6 and Eq. 3.3.3.7, the following equations are formed respectively;

Eq. 3.3.3.8 $E_t = (mc^2e^2\cos^2\theta)e^2\cos^2B + [(k)i^2\sin^2\theta]i^2\sin^2B$, and

Eq. 3.3.3.9 $E_t = (mc^2e^2\cos^2\theta)e^2\cos^2C + (\gamma c^2i^2\sin^2\theta)i^2\sin^2C;$

but, if we use the cross products of Eq. 3.3.3.6 and Eq. 3.3.3.7 the following equations result respectively;

Eq. 3.3.3.10 $E_t = (\gamma c^2i^2\sin^2\theta)e^2\cos^2B + [(mar)e^2\cos^2\theta]i^2\sin^2B$, and

Eq. 3.3.3.11 $E_t = [(k)i^2\sin^2\theta]e^2\cos^2C + [(mar)e^2\cos^2\theta]i^2\sin^2C.$

The validity to use both the dot products and cross products, and assume transmutation between these levels of freedom, stems from the ***observable natural phenomena*** that transmutation can take place between any of the four forms of energy above. However, the cross product transmutation will always end up mutually perpendicular to the dot (inner) products–thus giving three dimension to space. These transmutations do not preclude that latent energies will be needed to breakup/build closed systems; thus Newtonian physics may appear of non–effect during part or all of an energy transform.

As Eq.3.3.3.6 indicates the *mass function* $(mc^2e^2cos^2\theta)e^2cos^2B$ and the *active gravity* $(\gamma c^2 i^2 sin^2\theta)e^2cos^2B$ can exist simultaneously only as the dot product – cross product respectively at the B level of freedom. It is postulated that at the θ level of freedom $\gamma c^2 i^2 sin^2\theta$ exists but does not *appear* in Newton's relative frame, and that gravity exists exterior to mass objects. Proving this may be very difficult, but intuitively this makes more sense to me. We should account for the rigidity of spatial aggregates (galaxies, solar systems, objects, atoms, etc…) by an etheral fabric – I postulate that it is gravity energy.

Eq. 3.3.3.8 indicates why nearly 100% of our solar system's internal mass is in the sun with nearly 100% of its internal kinetic energy in the planets; furthermore Eq. 3.3.3.8 indicates our solar system's *exterior* gravity and absolute inertia may not be identifiable by earthly experiment, but equalling about 100% of our solar system's *interior* E_t to maintain equilibrium between its internal and external environments. From an external transcended frame we will only perceive about 50% of our solar system's E_t in the form of internal mass function and kinetic energy. Does this mean the absolute E_t of our solar system is roughly four times the mass function within it? Perhaps. It does not seem unreasonable to me to tie our solar system's motion through space to exterior gravity flux which causes its exterior inertia–entropy.

Let us approach life from a more godly point of view, after all God is the author of all life. About two–thousand years ago our Christ (who is symbolic of a meta–corporeal being) expanded our image of God to be a mystical Being of Three Persons in one Spirit. Likewise, we are made in His image: body, mind and spirit, or we are corporeal, meta–corporeal and incorporeal all in one mystical being. Let us derive a unitive equation based on these three unit vectors as parametric dual functions within themselves.

Let us assume that gravity is the source of all matter or substance; we then form a corporeal unit vector as follows:

Eq. 3.3.3.12 $(1)^a = m^2 e^2 \cos^2 a + g^2 i^2 \sin^2 a$,

where m^2 = Newton's mass = $(\gamma_{internal})^2$ or stationary energy, $g^2 = \gamma$, or the active gravity external to the mass in our closed system, and e and i the life vectors between these substances.

Similarly, our meta–corporeal unit vector can be written:

Eq. 3.3.3.13 $(1)^b = d^2 e^2 \cos^2 b + t^2 i^2 \sin^2 b$,

where d^2 = distance^2, or t^2 = (time)^2 = $1 - d^2$; and our incorporeal or unit life vector:

Eq. 3.3.3.14 $(1)^\theta = e^2 \cos^2\theta + i^2 \sin^2\theta$.

This life vector expresses the communion between the corporeal and the meta–corporeal.

The dot (inner) product of these three unitive vectors results in a composite unitive vector we call E_t, where:

Eq. 3.3.3.15 $E_t = (1)^a(1)^b(1)^\theta =$
$(m^2 e^2 \cos^2 a)(d^2 e^2 \cos^2 b)(e^2 \cos^2\theta) +$
$(g^2 i^2 \sin^2 a)(t^2 i^2 \sin^2 b)(i^2 \sin^2\theta)$,

as a, b, and θ "hover" about $45°$, or E_t = *a closed system*;

Eq. 3.3.3.16 $E_t = m^2 d^2 e_t^2 \cos_t^2\theta + g^2 t^2 i_t^2 \sin_t^2\theta$,

where the subscript t refers to the total life vector complex. At $45°$;

$m^2 d^2 i^2$ roughly $= g^2 t^2 i^2$, or

$g^2 t^2 i^2 = m^2 d^2 e^2 = (m^2 d e^2) d$.

If we eliminate the incorporeal vectors which Newton did not use and replace m^2 with m (Newton's mass), upon dividing by t^2 we see Newton's energy equation falls out:

$E = g^2 = (md)d/t^2$.

Newton made this mistake, his energy should have been an $E/t^2 = (inertia\ of\ g^2) + (md)d/t^2$ format.

This is evident by his Second Law, $F = ma$, when a ball rests on the earth it has no relative "F" downward because $a = 0$, but what does it have? The system of the earth and the ball have an inertia couple of g^2. This inertia couple also exists when the ball is falling, but Newton only equated the kinetic energy of the ball (entropy) to E.

Now returning to Eq. 3.3.3.16 at **45°**:

$E_t = m^2 d^2 e^2 + g^2 t^2 i^2$,

as $g^2 t^2 i^2 \to 0$,

because $(g^2 t^2 i_t^2 sin_t^2 \theta \to 0)$,

$E_t / t^2 i^2 = m^2 d^2 e^2 / t^2 i^2$; limit of $d^2 / t^2 \to c^2$; $m^2 = m$,
$\qquad\qquad g^2 \to 0$
$\qquad\qquad t^2 \to 0$

eliminating the incorporeal vectors,

$E_t / t^2 = mc^2$, is the corrected Einstein equation.

Apparently Einstein believed that the inertial mass and gravitational mass were synonymous. It follows that making Newton's mistake gave Einstein:

$g^2 t^2 i^2 = m^2 d^2 e^2$; $g^2 i^2 = m^2 d^2 e^2 / t^2$; limit $d^2 / t^2 \to c^2$; $m^2 = m$.
$\qquad\qquad g^2 \to E$
$\qquad\qquad t^2 \to 0$

Eliminating the incorporeal vectors,

$E_t = mc^2$.

This error has to do with the (=) sign, and it is a sublime error. One gets a better understanding of this error by replacing the (=) sign with the ($\rightleftharpoons$) sign; thus both equations read:

E transmutes to (force x distance), and vis–a–vis and; E transmutes to (mc^2), and vis–a–vis. With this approach, we identify E as a real substance, and not a derived mathematical function.

Newton and Einstein may have inherently had this in mind, but had they put a pen to it, the Unitive Equation would have been derived long ago by them.

This error is really a disregard for existence. If you don't believe that, convince your right hand that its existence is identical to your left hand's existence, after all your mind is a transcended impartial third party.

Another way of looking at this error is this: both Newton's and Einstein's equations describe a process, and time exterior to the process must be preserved such that *you* ($\rightleftharpoons$) and the *process–as–a–unit* enjoy your relativity, or *(ti)* dot *(ti)* = $t^2 i^2$, or conservation of the corporeal, meta–corporeal and incorporeal take place. Suppose we fuse two deuterium atoms into a helium atom such that $d(mc^2) = E$, before the reaction $d(mc^2)$ = *[itself, or unity (1)]* and $E = 0$, after the reaction $d(mc^2) = 0$, and E = *[itself, or unity (1)]*. The (=) now means scalar equivalence and, spiritual identity; however, during the process we don't really know how much $d(mc^2)$ or E we have. In order to maintain spiritual identity we have to go to a transcended frame (ourselves) where a new identity takes place as the communion of both $d(mc^2)$ and E, or $E_t / t^2 i^2$ = $d(mc^2 e^2) \rightleftharpoons E$. Ponder this for awhile.

Death occurs when these three unit vectors can no longer exist in resonant harmony; however, death of this highest order communion need not spell death for each unit vector in its lower order identity. The Bible speaks of evil spirits being cast out of Legend's body. Apparently either through poor communion of his own spirit, a vacancy for spirits was created, or perhaps as in our Christ — His sinlessness (and perhaps Legend's) created a perfect vacuum for His brother's sin, which rids us of our sin (and perhaps Legend's neighbors).

"Judge not lest you be judged" may not *just* mean by the standards you put to others so shall they be put to you; it may also mean you may be judged for imputing your sins into your brother, or why is your lack of faith?; your guile?; etc... found in your brother; thus the mechanics for redemptive suffering. Did you ever have the experience of thinking evil about your brother, and he went ahead and did just as you thought? To the degree that you have the same inner spiritual harmony as your neighbor, communion will be at hand, and if you don't, resentment is at hand. Did you ever understand the physics of why you resent those who are most immune to your persuasions? Perhaps this is occurring now because my physics has a life vector and therein a mathematical premise for philosophy, and yours does not have that inner communion?

Wave Mechanics doesn't come out of this error unscathed either. Suppose you only looked for Bluebills during September, and next September you find our Bluebill heading south; convince me that our original Bluebill wasn't the one I ate last October, or convince me after having not seen him during a day, your next sighting is of our original bird. Death occurs; *to equate a probability does not equate spiritual identity*. (This is another reason why I choose to deal in objects.) In all probability, your probability of resighting our original Bluebill to establish his habitat is probably pretty small; even though I have no uncertainty, you have a probable habitat for Bluebills.

Note that B as calculated from the gravity equation (see page 30) nearly numerically equals Planck's "second radiation constant", or $hc/k = 1.441 \times 10^{-2}$ *meters* $°K$ with $B = 1.443 \times 10^{-2}$ *meters/sec^2*. Can we logically equate the difference in units, if we use the entropy equation, $dS = dQ/T$, $T = dQ/dS$? Another way to explain this is: temperature relates the (internal work being done)/(free energy available). This internal work represents the kinetic energy of the ball's internal constituents as we rob them of their mass-at-rest inertia; however, free energy is also used to overcome the gravity net-cancellations (inertia) therein. Much like our ball being thrown into the air – we first must overcome its weight, or "latent heat" must be supplied to overcome the binary gravity couple between the ball and the earth, only after this phenomenon takes place can we impart motion to the ball to rob it of its mass-at-rest inertia. Additional free energy will impart

Newton's linear inertia to the ball (kinetic energy or entropy), or the ball's [mass-at-rest inertia + linear inertia (entropy)] is a conservative constant. As the ball continues upward its linear inertia will be overcome by the binary gravity couple and with time the ball will resume its inertia as it comes back to earth dissipating the first free and then latent energy we provided it. In this manner we can say the gravity couple's inertia represents negative entropy; thus the negative sign in my gravity equation. In Newtonian language,

$$F_g = \frac{\sqrt{G}\ m_e cos(0°)}{R} \cdot \frac{\sqrt{G}\ m_b cos(180°)}{R}\ , \text{or}$$

$$F_g = -\frac{G\ m_e\ m_b}{R^2}.$$

The negative sign representing internal inertia, or negative temperature (an isothermal heat sink).

Because Newton's force equation and Planck's "second radiation constant" do not take into account that the events they each studied, were accelerating around the earth's axis, as well as orbiting around the sun, we multiply Planck's "second radiation constant" by a unit acceleration vector; *hc/k meter (joules/joules) meter/sec²*. A unit acceleration vector is used here to preserve the values of Newton's and Planck's constants. The time for both of these events is an identity to the time in our unit acceleration vector, but in both Newton's and Planck's mathematical formats the spatial shift is not accounted for; thus in formatting the gravity equation (in Newtonian format *– f = Gmm/R²*), we must divide Planck's "second radiation constant" by the spatial shift he did not account for, or *hc/k = 1.441 × 10⁻² meters/sec²* (per meter). In light of the Unitive Equation the correct units for both Newton's force equation and Planck's "second radiation constant" are *g²meter²/sec²* where *g²* represents the corporeal substance of the gravity energy in the binary couple, or *g² = (the corporeal vector in Newton's "scalar" mass)* (more on this later). If the logic of this scenario is accepted, then we have a direct relationship between gravity energy and light energy.

51

To approach our former problem where the high frequency photons of white light are refracted more by a prism, let us list some historical equations we use to study mass-energy transmutations and/or resonances:

$E = (Gmm/R^2)(x)$ (Newton's);

$E = mc^2$ (Einstein's);

$E = hv$ (Planck's); and

$\lambda mv = h$ (de Broglie's).

Furthermore, c can be partitioned into two inverse variables, $c = \lambda v$.

We might ask: why do we need the Unitive Equation ($E_t = mc^2 cos^2 \theta + hv \sin^2 \theta$) when we have all these others? The answer to this lies in the perspective of each founder's logic. Physicists deal in constants, scalar quantities and vector quantities. If we raise each of these equations to an event-basis where substance is represented by internal mass function or external field function, where space and time become a continuum, and the event-basis constitutes existence, we see a pattern emerge from these equations:

Newton's equation $E = (Gmm/R^2)(x)$ scalarizes substance and allows E to vary as distance/time and existence (the location of the event) vary;

Einstein's equation $E = mc^2$ scalarizes the event c and more specifically distance of the event c and allows E to vary as substance/time and existence vary; and

The Unitive Equation scalarizes existence and allows E_t to vary as substance and distance/time vary.

Note: if we unitize E_t such that it also becomes a scalar, then the other remaining variables, vary either inversely to one another as in $c = \lambda v$, or we can scalarize any number of the variables, but we must leave at least one pair of variables inverse to one another to account for all intra variations between E_t's internal and external components as E_t responds to a changing environment.

Before I get to how this pertains to Planck's and de Broglie's equations, I want to once again explain why scalarizing the existence parameter is necessary to solve life problems.

We base our time parameter on the physical phenomena of our earth in motion, and one second of time represents the time elapsed during a certain meridial deflection of our earth about its axis. Let us say one second equals the time elapsed during a deflection angle θ. If I want to study your kinetic energy while you are on the earth and while I am on a large star which is slowly rotating (such that the star's same angular deflection, θ, represents one second to me), we first must equate our time. Let us say one second to me represents 10 years of seconds to you. Based on my time, if I were to write a Newtonian equation for your kinetic energy such that $K = 1/2\ mv^2$, I would end up with an absurdity because your mass at the start of my second versus your mass at the end of my second is not an identity. The only way I can write this equation is to subordinate your mass as a variable to your existence or life vector, i.e.; I must scalarize your existence vector, in that way I can say your existence vector consists of these molecules at the start of my second, and your existence vector consists of these different molecules (which you ingested during your ten years) at the end of my second. It becomes apparent that if I am on the earth this same phenomenon (change in mass identity) is constantly occurring in every object but to a lesser degree.

This same phenomenon is true if I were to scalarize our distances traveled in Einstein's relativity equation.

In order to study life problems we must scalarize or not vary our existence vector; in this way, "you" represents "you" no matter how your other parameters vary. Physicists do this, but I don't believe they know they are doing it. Example: If we are studying a problem involving a spring - we give the spring a certain elastic constant - this scalarizes the spring's existence because throughout the spring's life, its elastic constant will vary. "You", meaning the spring, only pertains to the spring as it exists with its present elastic constant.

This phenomenon of scalarizing (fixing) a certain parameter of an event to understand or perceive the other parameters of the same event has led the author to make the following statement:

We perceive the existence of an event by that which is not of the event, i.e, perception is a binary process of emanation of something from the event which is received by our perceptive senses. This modifies the philosopher's statement "that things only exist in the mind" to its corollary – that the mind only exists if it can perceive emanations from other existences. Incidentally, this would indicate that all objects have minds because all objects have self seeking gravity fields capable of metaphysical perception and simultaneous binary physical reaction with gravity fields emanating from other objects, or all objects have the basis of life inherently within.

Examples:

1. You do not see the existence of this page, rather you see light which emanates from it;

2. You do not taste food, you taste what it gives off;

3. You do not touch this book, rather you are extracting some of its inertia and/or kinetic energy of its former state of existence;

4. Except for changing its mass-at-rest inertia to linear inertia, you do not put energy into lifting its mass, you put energy into separating the book's gravity field from the earth's; and

5. Though you might "smell" a skunk in the physics patch at this time, you can only "smell" what I emanate.

On an historical note, it should be recognized that there is a tendency to scalarize that which is most tangible to us. Newton scalarized mass (the object's corporeal); I believe that Einstein's uncertainty of E, m, and t caused him to scalarize distance (the object's meta-corporeal), and my certainty causes me to scalarize existence (the object's incorporeal). It should also be recognized that we have a tendency to make constants of the mystical unseen events that combine the above. *E, G, h* etc., are *real* unseen physical-metaphysical *events* in nature of which we only see the results; so, in our ignorance we assign numerical constants to these events.

In the former problem where I calculated $g_{lat.} = Gm/R^2_{lat.} - B\sin^2_{lat.}$, I scalarized *R* for each latitude by precalculation because my logic tells me that the geometric shape of the earth is an expression of the earth's internal gravity (inertia) versus its kinetic energy (entropy). Keep in mind *R* does vary over the domain of the equation, but by scalarizing *R* at each latitude, I place the total differential effect into *m* because *G* remains constant throughout the equations domain. I could have varied *g* simultaneously with *m* by $G_{lat.} = GdG$ (where $dG = \cos^2\theta + \sin^2\theta$ function), but my radii of the earth were calculated from a theoretical ellipse, $x^2/R_e^2 + y^2/R_p^2 = 1$, and I would not attempt doing this until the empirical radii were introduced into the equation.

How does the above pertain to Planck's and de Broglie's equations? Let us first look at de Broglie's equation, $\lambda = h/mv$ as it pertains to an electron with a kinetic energy of 100 electron volts. Based on this scalarized kinetic energy of the electron, and the scalarized mass of the electron, we can calculate how distance varies with time by the equation $k = 1/2mv^2$; $v = (2k/m)^{1/2}$, substituting, $v = (2k/m)^{1/2}$, into de Broglie's equation yields: $\lambda = h/m(2k/m)^{1/2} = 1.2$ *Angstroms*.

What is the meaning of this λ we have calculated? Because we have based λ on only the kinetic energy (that which was emanated to us), λ in turn is only descript of the energy emanated to us. (Where did this kinetic energy come from? Did it come from a change in the linear velocity of the electron? Did it come from a change in the rotational velocity of the electron: Did it come from a change in mass function of the electron? Or, did it come from a change in all three of the above?) When we scalarized the linear kinetic energy of the electron as 100 electron volts, we presumed that the mass and the rotational velocity of the electron did not change during the emanation of the kinetic energy. In doing this we also presume that the "electro-magnetic" wave of the kinetic energy emanated had zero rest mass because our electron's mass function did not change.

We might also ask is **h** a constant? Because $h = \lambda mv$, we might conclude that **h** varies with λ because we can scalarize **m** and **v**. This may have meaning when we apply de Broglie's equation to a photon or when $v \to c$; such that $c = \lambda v$, where **c** is a unitized inverse relationship between λ and v.

$E = hv$ (Planck's Equation)

$h = \lambda mc$ (de Broglie's Equation)

$E = (\lambda mc)v = mc^2$

Allowing **h** to vary, to "float," is no different than allowing **G** to "float" in our former gravity equation at 0° latitude. Another way of expressing this is that in $E = hv$, **v** reflects the *internal* stability (inertia) that maintains the photon as an entity during transition, while **h** reflects the *external* free energy (entropy) of the photon both in inverse relationship as we unitize **E**. The author believes the Unitive equation expresses this better by partitioning inertia from entropy, or

$E_t = mc^2 cos^2\theta + hv\, sin^2\theta$.

This scalarizes the existence of E_t and allows the inertia and entropy to vary reciprocally, but it also leads us to some fundamental questions:

1. Can $hv\, sin^2\theta$ vary without a variance in $mc^2 sin^2\theta$ as we vary E_t? (This is what we presumed when we scalarized the kinetic energy of our electron);

2. If E_t is split such that two new E_t's are generated will both have internal and external stability? (This may give us some insight to the "string phenomenon" or "string theory".)

3. Upon splitting E_t, does $mc^2 cos^2\theta$ stay behind while $E = hv$ is emanated to us? Remember we perceive existence by what is emanated to us, so *does Planck's equation represent only what is emanated to us*, or does it represent the total energy of our original E_t? (In our electron problem we presumed $E = hv$ represented what was emanated to us because our electron's mass function stayed behind.)

The author believes that any time E_t is split: two or more Unitive equations will represent the split factions. The existence of the two or more entities would seem to demand this. Another way of saying this is: if you can locate "existence" then you will simultaneously locate internal and external stability.

In de Broglie's equation $\lambda mc = h$ at c, m is indicative of the photon's internal stability. (Whether you view the photon's internal stability as a tiny particle of mass, an internal net-cancellation of gravity, or an internal net-cancellation of an electro-magnetic complex is an irrelevant point here – the existence of the photon gives it an internal stability which is represented in de Broglie's equation by Newton's m.)

As white light is refracted by a prism, de Broglie's equation predicts that if we scalarize h and unitize the event c, then λ and m will vary inversely as E_t varies, or as λ increases, m decreases and as m increases λ decreases; thus there is a direct relationship between the internal stability m and the frequency of the photon, v. As m increase so does v.

If we graphically present these photons as sinusoidal waves traveling at c, we see that the curve lengths of the higher frequency photons are much longer per the linear event c. The author refers to this curve length as the photon's wave front phenomenon, and there is a direct relationship between this wave front phenomenon and the internal stability of the photon. It would appear that because all photons are unitized to the event c, the higher energy photons have an increased secondary motion to account for this increase in wave front phenomenon. [The only way both the event c and the wave front phenomenon (curve length) can be unitized over a range of photons is for the effective gravity radii of the larger photons to be proportionately smaller, which seems opposite to Newtonian law – $f = Gmm/R^2$. Yes, we can see that as R decreases, f increases, but this means that the increased stability of the photon in question is based on exterior consumption of its gravity, rather than interior net-cancellation of its gravity. Even though our sun's stability is in part due to exterior consumption of its gravity field by our planets, its effective radius for planetary motions seems proportionate to its internal stability, or its mass.] This has lead the author to the following statement.

"Charge" with its electro-magnetic characteristics is nothing more than the secondary effects of a gravity field being sheared or in motion, and the exterior consumption of an objects gravity field provides stability to its kinetic energy (entropy).

The majority of planets in our solar system utilize their gravity fields in two ways: 1. to maintain their solar orbits; and 2. to maintain their satellite's orbits. If we look at the "relative charge" that a planet can use for its satellite system versus its "charge" used to maintain its orbital motion, based on a Newtonian format

$$\text{"Relative charge"} = \frac{\text{kinetic energy of rotation}}{\text{kinetic energy of orbit}}.$$

"Relative charge" $= K_r / K_o = 1/2\, I_r W_r^2 / 1/2\, I_o W_o^2$,

"Relative charge" $= 2/5\, mR_r^2 (2\pi / T_r)^2 / mR_o^2 (2\pi / T_o)^2$,

"Relative charge" $= 2/5\, (R_r^2 T_o^2 / R_o^2 T_r^2)$.

Table 3-2 lists the ratio of each planet with its number of satellites. (Saturn's rings are not included, and more space probes may increase the satellites listed. Pluto's data was omitted because its sidereal day was not found.)

Table 3-2:

Planet	Ratio of K_r / K_o	No. of Satellites
Mercury	1.68×10^{-9}	0
Venus	7.39×10^{-8}	0
Earth	9.76×10^{-5}	1
Mars	4.07×10^{-5}	2
Jupiter	0.35	16
Saturn	0.38	17
Uranus	0.15	15
Neptune	0.10	6
Pluto	?	?

Table 3-2: Ratio of planets' charge with number of satellites.

There is a definite trend between the planet's "relative charge" and the number of satellites each planet will maintain. Further consideration must be given to the total kinetic energy of each planet's aggregated satellite system to be able to formulate, or elevate "relative charge" to a "universal charge" formula. Other than the earth's moon, I have no data on the mass, orbital radius, and period for the other satellites listed so that the total kinetic energy of the other planet's satellite systems might be aggregated. Jupiter's "relative charge" ratio is slightly less than Saturn's, but Jupiter's "total charge" would seem a function of its mass which is far greater than Saturn's. We could leave m out of the denominator and write an equation as follows:

$$\text{"Total charge"} = \frac{\text{kinetic energy of rotation}}{(orb.\ accel.)\ \times\ [d(x)\ of\ the\ orb.\ accel.]},$$

or "total charge" = a mass function such as $g^2 i^2 \sin^2 \theta$.

It seems pertinent to leave such formulation to the future, after our astrophysicists have had time to digest the significance of the "relative charge" formula. The most that could be speculated at this time is that a non-rotating body cannot maintain a satellite complex because it has no "relative charge", and that if a sun were to eject a planet from its polar region (with similar axial alignments), there would be a potential to spiral the planet toward the sun's equatorial plane (align or net-cancel the planet's "charge" with the sun's "charge" for internal stability); thus, making a disk-like solar system. Because the axial alignment of the sun is viewed as inertial alignment (magnetism), the first event in planet ejection would be to break down this inertia, with "charge" (entropy) maintaining our solar system's internal stability by pulling the ejected planet to the sun's equatorial region. Do you see how this scenario of multiplicity of gravity motion equates to spin function kinetics?

[Pertaining to electric circuitry, I find it a bit absurd that the classical physicist develops a variable *emf* from electrostatic charge, but the charge remains invariant during the electron transfer. If charge is a gravity field in motion, we could say that the chemical energy in the voltaic cell causes an inertial pole by robbing "electrons" from it, while the other pole gathers these excited (faster moving objects with gravity fields) "electrons" causing an entropic pole. As the "electron" moves through the circuit it gives up some of its entropy to the substance (mass or energy field) of the circuit, or its charge decreases as it moves to the inertial pole.]

How does this relate to our prism problem? If it is true that the internal stability (de Broglie's–Newton's mass) of the photon is proportional to its frequency, then the effective gravity radii of the higher frequency photons will be greater. These higher frequency photons will begin to bend toward the prism sooner than the lower frequency photons as our beam of white light approaches the prism's surface; thus more time is consumed, and more bending occurs for these higher frequency photons. It is also reasonable to equate these larger effective gravity radii to slower photon velocities while in the prism because these larger effective radii will interact with more mass of the prism media. (In Einstein's Theory of Relativity the secondary effects of the "relative charge" will have more relative time to interact per linear event c.) Though not convinced, the author suspects the secondary motion of photons to account for their variations in wave front phenomena is as our planets; it is a rotation of the photon about its axis. How this axis is spatially oriented with the linear event c is another matter? A question keeps popping into my mind that I want to rid myself of here. If the photon is the *fundamental building block* of all matter, do all photons quantitatively have the same E_t with the amount of energy emanated to us being a function of how their secondary wave front phenomena are aligned to each photon's linear event c? Or are photons like planets? Do they have the ability to be unitized to the event h as their E_t's quantitatively vary, just as our planets are unitized to the event G, with both the photon's and the planet's "relative charge" being a secondary response to h and G respectively? The later would seem to be the case because the author believes that just as in our $g_{lat.}$ problem, we can show that h is only a specific entropy expression of the inertia event G; however, the above question of axial spatial alignment, due to the fundamental nature of the photon, needs to be asked.

Given: $F = Gmm/R^2$.

If we unitize the gravity event such that $m \cdot m = 1$ and $R \cdot R = 1$, with the understanding that the geometry of the problem must be depicted by the denominator, we can alter Newton's point-to-point expression by spreading the G event over the area of a unit sphere.

$$F = \frac{(6.67 \times 10^{-8} \, dyne \, cm^2/gr^2)(1gr^2)}{4\pi(1cm^2)} = \frac{6.67 \times 10^{-8} \, dyne}{4\pi},$$

$F = 5.3078 \times 10^{-9} = a_0 =$ the Bohr radius expressed as dynes.

It follows: $G = 4\pi a_0 = \dfrac{4\pi h^2 \in_0}{\pi m_e e^2} = \dfrac{4 h^2 \in_0}{m_e e^2}$.

Here is proof that **charge** is a **gravity** event. I honor Newton here because by this same equivalence - gravity is a charge event. We must also honor all the other physicists of the past because this expression for **G** unites all of their efforts into Newton's minimum set. If you are concerned about the units of dynes consider:

Gravity Energy $= F \cdot a_0$ where a_0 is the relative radius.

Consider that Einstein and others did not feel the need to equate charge phenomena to "fundamentalism," and that the planetary problem just discussed provides a glimpse at how gravity is used for multiple stable resonances within resonances, do you see why **motion** not **charge** is fundamental, and that as we traverse our universe from nuclear order through galactic order the various motions (relative spin functions) are indicative of how gravity is being consumed at each resonant level?

The following experiment is suggested to determine what phase angle exists in the Unitive Equation for various frequency photons.

It has been known for sometime that photons exhibit mass properties (momentum) when light is allowed to fall on a mirror suspended by a vertical quartz fiber. The following *suggested experiment*, illustrated in **Figure 3–4**, utilizes a similar quartz fiber in a horizontal mode.

Figure 3–4:

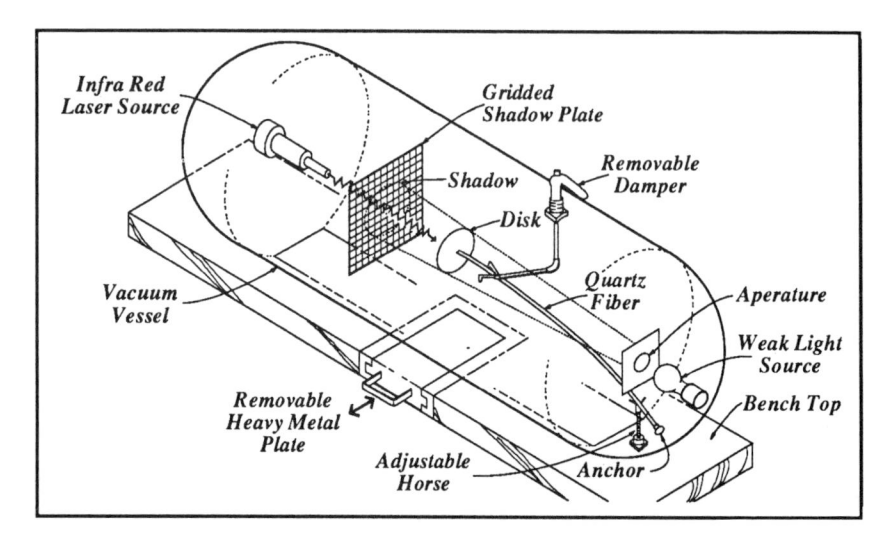

Figure 3–4: Suggested Experiment

    The quartz fiber should be calibrated and anchored such that the disk (which is mounted perpendicular to the fiber) lies in a vertical plane when the damper is in place. The plane of the disk should also be perpendicular to the laser beam. An adjustable horse may be needed for proper alignment of the fiber and disk. (It may prove more expedient to hang the disk from the quartz fiber rather than mount the disk perpendicular on to the fiber.) The physical properties of the disk should be that it can absorb infrared radiation and undergo thermal expansion. The damper should be placed as close to the disk as possible. The purpose of the damper is two fold: 1. It should prevent any resonance or strain to the fiber while the laser strikes the disk; 2. It should also act as a heat sink to prevent the quartz fiber from heating up so that its elastic modules will not vary. The damper should be raised a constant distance above the fiber when not in place to prevent any variation in gravitational attraction between the fiber and damper. The purpose of the weak-light source and perhaps aperture (if needed) is to provide a shadow on the gridded shadow plate. Prior to experimentation the disk should be put through a number of temperature variations with the weak-light passing by it to determine if the refractive nature of the disk varies with temperature. If so, proper corrections should be applied to the shadow upon experimentation. An infrared monitor should be employed to determine the temperature of the disk and fiber during experimentation with proper corrections applied.

Experiment:

The damper should be put into place and removed several times to determine if the fiber gives consistent results prior to the laser being turned on. If a heavy metal need be used to increase the Newtonian *g* factor on the disk, then it too should be both in place when the damper is being put through its dry runs, and also absent with both positions of the shadow noted.

Remove the heavy metal and place the fiber in position with the damper. Allow the laser to pass through the hole in the gridded shadow plate striking the disk until the infrared monitor indicates that the disk is radiating as much energy as it is receiving. Turn the laser off and put the heavy metal plate in place while simultaneously removing the damper.

If the disk gained thermal energy - the shadow should expand (allow for refractive correction due to temperature).

If the disk simultaneously gained mass - the shadow should appear lower on the gridded shadow plate.

If a simultaneous mass increasement has been observed, the disk should be allowed to radiate its excess thermal energy to the former position of the shadow when the heavy metal was in place prior to turning on the laser. If the disk does not obtain its former position this may indicate that the increased Newtonian *g* factor has altered the ability of the disk to radiate thermal energy.

Remove heavy metal and allow the shadow to return to its former position when heavy metal was not in place.

Using a different infrared frequency repeat the above experiment.

Note: Allowing the heavy metal to be in place while the laser is turned on may alter the phase relationship of the incoming infrared photons. Though not known, the author suspects that a simultaneous mass increasement will accompany the thermal expansion of the disk. If so, a phase relationship should be calculable between Planck's free energy $E = h\nu$ and de Broglie's–Newton's photon mass $m = h/\lambda c$. This in turn will determine θ in the Unitive Equation ($E_t = mc^2 \cos^2\theta + h\nu \sin^2\theta$). If no mass increasement is observed it means θ is approaching $90°$.

Different disk materials should also be used: hydrocarbons; graphite; mica; platinum; etc....because the disk's interior media may alter the phase angle of the absorbed photons.

3.3.3.10 The Speed of Light

What does the speed of light mean in unitive thought? It means the velocity of all interactions, $c^2 = d^2/t^2$, or the maximum velocity that can take place at any instance of time. For a free falling body c maximum equals roughly *121 mph*. Remember everything is made of gravity interacting at c. Newton's mass *internally* equates to $m_n = (\gamma c)(\gamma c)$, or $\gamma^2 c^2$, but externally $\gamma^2 c^2$ becomes m, and $(d^2/t^2) = 121$ *mph*. Why does the square $d^2/t^2 = d/t$? Collectively both the earth and the falling object are experiencing the same event. (The *121 mph* is c^2 in our transcended reference frame; however, the classical physicist will impute the event to only the ball in his *internal* reference frame as he observes the ball falling relative to himself.)

Does the limit $d^2/t^2 \rightarrow$ infinity, as $t \rightarrow 0$? I don't know. If $t = 0$ it could be argued that existence goes out-of-existence, perhaps the time vector exists but rotates $90°$ away from the gravity vector. In Unitive thought c is a real event and carries equivalence, as well as spiritual identity. For photon transfers relative to other objects, c can go to infinity if we are only observing c's wave front phenomena which are real events. [The author has never seen a *black hole* or a *quasar* but I would like to explain two events that are consistent with this Unitive Theory. Suppose you are viewing a large gravity centroid out in space (a *black hole* in Wheeler's terminology) that suddenly disappears, but a short while later another *black hole* appears somewhere else in the cosmos. How could this happen? The author believes that all objects or closed systems have a certain stability about themselves which has to be overcome prior to destruction of the closed system. The reader might liken this to *latent heat* phenomenon, or the *probability* of Wave Mechanics. Our *black hole* will be stable as long as the external need for gravity is balanced with its internal need for gravity; however, if other large galactic systems are moving about it, and their kinetic energies allow them to come close enough such that the stability of the *black hole* is overcome; a new *black hole* will form at the center of gravity of these large systems at the expense of the old *black hole*. Keep in mind that the center of gravity was always at the new location, but the *black hole* phenomenon represents — the location of the centroid where the gravity impingement is most stable. Such an event may appear instantaneous. This phenomenon may be accompanied by a cosmic *aurora borealis* when these wave fronts realign their stability — perhaps this is a *quasar* phenomenon.]

The Speed of Light Continued:

As shown on page 61, all of the modern day physic constants are tied to its grandfather's (Newton's) definitions of G. Is G universal a quasi-constant that varies proportionately to the life vector's attitude. We can pick up our foot and overcome gravity, but internally do our molecules function differently because internally G is different? Similarly does the velocity of light in the nuclei of atoms vary due to how diffused their masses are? In the past, fission and fusion potentials have been equated to excessive mass in the nuclei of atoms, but if $m_n = \gamma^2 c^2$, does γ vary with G to give rise to fission and fusion potentials due to variations between c in these nuclei and c in their exterior space?

3.3.3.11 The Future

Before looking to the future, perhaps we should review the past. If the author were to choose three fields of endeavor to encapsulate mankind's past achievements, they would be:

1. Theology;

2. Physics; and
3. Philosophy.

These three "sciences" would seem to be a mirror image of the Holy Trinity and the Unitive Equation.

The theologian subordinates both the material and the mental-emotional complex to seek continuity with the mystical or spiritual resource of the universe. He nevertheless relies on physical manifestation of his prophecy and prayer to justify the logic (mental order) of his endeavors;

The physicist subordinates the spiritual and mental-emotional complex to seek the interaction of the visible substance (material order) of the universe. He nevertheless relies on the faith of his assumptions to justify the logic (formulations) of his endeavors; and

The philosopher subordinates the spiritual and physical complex to seek a logical format (mental reasoning) to discuss what he senses. He nevertheless uses both gray matter (physical order) and its communion with life's energy (spiritual order) to justify the existence and lack of existence of his endeavors.

Once again, we can see a pattern emerge from the above. If we place our certainty (scalarize) one of the above, we destine our conclusions.

The Unitive Equation is a communion between each of the above endeavors, scalarizing the Theological entity into *n* roots. The purpose of scalarizing your existence rests in the need that now you as an object can suffer *n* roots of substance and time-space transformations and still remain you until your object no longer has the ability to exist.

The Unitive Equation with its multi–dimensions/multi–level life vector hopefully will provide science with an inchoate mathematical system for future exposure of the Trinity in nature. Remember God spoke to earth while His Son was on earth, and Christ imparted the Holy Spirit onto the Apostles; similarly the equation's three entities may exist simultaneously separate, as well as unitively. I see many mathematical explanations in this equation for formerly thought delusions.

Let $E_t = g_i^2 d^2 e^2 \cos^2 a \cos^2 b \cos^2 \theta + g_e^2 t^2 i^2 \sin^2 a \sin^2 b \sin^2 \theta$ with both intra–life within the corporeal and meta-corporeal vectors, as well as, communal life between these vectors system.

What happens when $g_i^2 \cos^2 a \to 0$, as $t^2 \sin^2 b \to 0$, with $e^2 \cos^2 \theta$ remaining at $45°$ – Do we have a "ghost" traveling at the speed of light?

Is death when our life vector rotates $90°$ away from our inertia/entropy vector with our meta–corporeal vector being mutually perpendicular to both such that each resides as a dot product to an exterior vector?

What of related cross product phenomena between outside E_t's? Does this lead to higher order dot product phenomena?

One of the chief reasons why the author was motivated to contemplate physics is that equations such as $E = mc^2$, $f = mg$ violate a fourth grade principle of mathematics, i.e., *one orange = (one apple) (bananas)^2*, and even when we raise this system of understanding to vectorial mathematics, lack of definitive continuity from one side of the equation to the other is missing. In other words we need to express more than equivalence; we need to express identity; and in that light, the author feels that he has taken the rudiments of his existence and has successfully with the Unitive Equation depicted a state of awareness — that there are no glitches between classical, thermodynamic, quantum or relativistic physics, that these lack of continuities are inherent to the system of mathematics used to examine physical behavior and more specifically to nearsighted (laboratory) presumptions. The reader may not agree with the author that gravity fields have self-motivating (metaphysical) properties, but he will nevertheless have to rely on mysticism to explain gravity behaviors. Newton's system of equations deals with one level of freedom (mass internal – energy internal) which can be applied independently to n systems. Einstein's system depicts light transfers do not need Newton's internal constituents to exist; after all, light energy can transcend several levels of freedom. Light can be emitted from a subatomic particle and travel to another galaxy. So Einstein let the characteristics of light predominate as a constant, and we

confirm Einstein's postulation based on laboratory experimentation that "apparently" proved that light needs no media to exist. And indeed Einstein's equivalence does heighten our awareness of this freedom from Newton's Laws; however, Einstein's mathematical system breaks down when he postulates that dense gravity fields will effect changes in a photon's behavior; and in fact, these postulations are observable. This breakdown is caused by Einstein's reliance on Newton's internal reactions. Photons need their own gravity fields to interact with these other exterior gravity fields for effective changes in their transfers to take place. If this were not so, photons (by the authority of Einstein's mathematical system) would be immune or not affected by gravity fields. Two questions arise from this:
1. Did Einstein's postulation that light needs no media for propagation violate the authority of his mathematical system? No! His mathematical system is actually in violation of his postulation. If he in fact postulates that his photon has a gravity field that interacts with his dense exterior gravity field to effect a change in the photon's behavior; then too, he should have accounted for the effect of this photon's gravity field interacting with a less dense exterior gravity field. His system must show that definitive continuity persists between these two environmental states of his photon; and 2. Which exterior gravity field caused this effective change to his photon? Is it a field from a micro constituent of the system? Is it a field constituent of the aggregate of the system? Is it a field imposed on the aggregate system from the macro environment? The answer to this question will lead one to a *physical* riddle unless you recognize that gravity, like Einstein's light, has the Unitive property that it can also transcend all levels of freedom (or all levels of resonant systems) within the universe. (Mass can transcend these levels of freedom only with the aid of an expendable source of internal energy; however, light transcends all levels of freedom with its own non-expendable energy within our universe and if light need a media – gravity is always available for interaction. With these states of compatibility of freedoms, the author contends that gravity is the media for all "electromotive" transfers. A photon is a Unitive form of gravity in a media of gravity – like all other objects.

A paradox of the Theory of Relativity is that it derives the concept that there are no absolute motions based on the absolute motion of light. This paradox apparently exists in Einstein's "time-warps". It is caused by placing an inertial frame in motion or trying to elevate perceived behavior to absolute behavior (giving *c* dominant anthropomorphic characteristics). In essence it is a violation of mathematics – a photon's time, mass, behavior etc., cannot alter my time which is based on the chronology of events hrs./days/years; i.e., I cannot perceive a "time-warp" unless I have the mathematical authority to say that my reference frame did not alter its time during the warp, and by that same authority – the photon cannot alter my sense of time. The corollary is that if I can perceive photon(s) as they are going through a "time warp", I should be able to formulate a set of equations describing this behavior in my own time reference frame; in doing that, the velocity of light will become a variable based on the authority of my time. Though his math may explain perceived "relative" behavior it places the anthropomorphism in the wrong set of Unitive parametrics (the meta-corporeal). We need to place them in the incorporeal set. One may argue that the incorporeal set is really, *(magnetism^2 + charge^2)*, but this sterilizes the equation or kills its anthropomorphism. I would rather take the bold step and introduce the terms *(life$_i$^{2} + life$_e$^{2})* to physics, and recognize that both Newton's and Einstein's systems represent special cases of the Unitive Equation. In short this mathematical violation is also present in Newton's "inertial" frame. Both Newton and Einstein subjugate the observed object's time and existence to each reference frame, rather than viewing both simultaneously. This becomes apparent in Newtonian physics because most physicists believe that *g* of *f = mg* is a vector when it is in fact the scalar product of two gravity vectors in opposition. This is why *g* may be calculated at the earth's surface, and synonymously *f(a scalar) = m(a scalar) g(a scalar)*, or the *f* calculated by this equation is inertia as previously mentioned. Now we have definitive continuity; when *(dx/dt^2) downward = 0*, we measure inertia and not force. A classical physicist would jump in here and say whoa! Inertia does not have the same units as force. I will leave it to the reader to contemplate that both the earth and the ball remain motionless to one another while enjoying the same higher order motion which in turn gives inertia higher order vectorship.

The author believes that the next major advancement of science needs metaphysical characteristics and that these metaphysical characteristics do in fact exist by their own set of laws which are in harmonic coexistence with the physical. We may envisage that the universe is constantly changing, but in the same stroke we have to realize that such a metaphysical phenomenon (this vision) is housed in, or emanates from a physical body, and that these coexistences of metaphysical/physical are harmonic. Once one realizes that such phenomena are real, it is difficult to impart only physical characteristics onto/into objects because you cannot create something from nothing per conservation of energy.

By the harmonic behavior of natural law it would not seem unreasonable to deduce that the physical nature of an object will determine its metaphysical characteristics and vis-a-vis that its metaphysical characteristics will determine its physical state. Religion, medicine, psychiatry have numerous documented experiences testifying to this dual relationship, and in fact all of science has this incorporated into it. All physical law formulations rest in the fact that physical behaviors are metaphysically experienced. Newton could not cut out a piece of free gravity field and take it into his laboratory and poke at it. Gravity receives identity through experience. It physically does exist, and is only perceivable when we utilize life to experience it; thus with a double scotch and a dash of humor all physical laws are subordinate to the metaphysics Christ spoke of. We need to join the harmonics of both systems to allow our sciences to continue. Over emphases of physical laws may skew our perspective and retard the growth of knowledge. The reader may find this more philosophical than scientific, but ask yourself – why don't I have a scientific procedure to prove faith, to prove hope or to prove love exists? We might, if we allow proof to come to us by other than physical vehicles. These states of grace (faith, hope and love) are infallibly conscious to our metaphysical person, but because we accept only physical evidence as vehicles of proof, these states of grace go unproven to our neighbor and to our doubting selves when we try to collect, or aggregate our grace. The author has attempted to bridge these two states of harmonics by allowing all physical objects to have life properties. In the classical sense this may be an absurdity, but realize that when we identify an object, say our sun, that identification imparts a mathematical authority to that object – no matter how its mass varies it will always be our sun and with that authority of existence comes life properties for coexistence. The concept that an object has a fixed amount of mass, without the inherent ability to internally/externally respond to its environment, is an irrational dichotomy which our sun disproves every day, as well as you presently using your daily bread to read this treatise.

We could also approach life utilizing the Unitive Equation as an integral: Life $= \sum_{1}^{n} E_t\ (n)$ where n represents those separable components that contribute to the whole object's existence.

In all these approaches there is room for sets of matrices to define what physical species of life can coexist with its environment, i.e., with each object comes a set of environmental conditions to maintain its existence (it has to maintain a certain density range, temperature range, fuel source or free energy for interaction, inertia/entropy ratio, motion, etc.). All these physical characteristics can be matrixed, and the variations within should lead us to expected cyclical life forms due to variations in these internal vs external parameters. Can human life exist when the earth's crust density is 10 gr/cm^3? Genesis 6:4 speaks of men of great renown, perhaps they had to be physically larger and stronger to be mobile when the earth was more dense? Similarly can an amoeba fall to the bottom of the ocean and maintain life or do the tremendous pressures there kill it? If it were to fall at a very slow rate, does it have the metaphysical resilience to physically ward off this state of death? I feel, we need to create and examine these matrices to further understand life. Not that I wish to impart souls (conscious awareness of God) into all things, but more – that a life parameter exists in all objects for sustenance of co-existence, and that the various physical forms of life go hand-in-hand with the metaphysical harmonics therein (which polices the physical and vis-a-vis).

Inchoate as this physical/metaphysical linkage appears there are reams of matrices available for research, understanding and eventually formulation of why physical objects house or emanate metaphysical characteristics to form life. I offer the Unitive Equation in its various formats to provide a conservative–parametric–oscillatory function with *n* roots of freedom to begin this new journey in science. I am sure there are others. While it is true that similar formulations could have been developed from electrical and magnetic fields being the fundamental building blocks of the universe, rather than using these bi-fundamental parameters, I concluded a mono-fundamental parameter (gravity) sufficient to format the physical substances of the universe with the mystical-nonphysical (metaphysical) parameter being fundamental to sustain the physical's oscillatory existence.

In my initial paper I introduced two variables to explain why parametric oscillations are natural – *inertial potential* and *entropic potential* are at the root of all dynamic motions. When I did this I was trying to explain why two equivalent "scalar" masses can have opposite potentials for mass growth, i.e., our earth is in a state of growth due to incorporation of heavenly meteorites and an equivalent amount of mass in our sun is in a state of decay. Upon more recent reflections, dynamic response of gravity forms to these two potentials may be the physical expression of the metaphysical portion of the universe. No matter how many physical laws we derive to explain physical behavior, at the root of all these laws is a metaphysical parameter of why is there a state of oscillatory or resonant behavior? Derived physical laws only predict physical behavior, they do not cause physical behavior, while absolute law unites both the cause and the behavior into one being. Thomas Aquinas tried to tell mankind this many years ago.

Recently I was studying Kepler's Law of how the sun's mass can be calculated without using the mass of a (the) planet(s). I was astonished to find that his law predicts that the sun's finite mass had the ability to support an infinite amount of satellites; because without Unitive properties to our solar system, that's what his mathematics states. This same law has a metaphysical characteristic when we consider our solar system is a subordinate unit within the galactic order of our galaxy. To the author – Kepler's Law states that if a satellite is going to orbit about the sun, its mass and kinetic energy will cause it to obey the Unitive harmonics within, which will be expressed by its orbital radius and period. Can you imagine what disorder would occur if G universal for the Sun–Jupiter harmonic was twice or three times greater than G universal for our Sun–Earth harmonics. It is the author's opinion that galactic order demands that our solar system act as a unit, with its mass–momentum or inertia–entropy characteristics quasi-constant in its subordination. And that its long term variations precess as galactic order precesses. Also, I am not a great believer in non-varying constants – constants are just tools to legalize relative physical behavior. When we can stand outside our universe and look into it, we may be able to form non-varying or absolute constants. And now to get back to our neophyte sun.

3.3.3.12 The Sun Continued

As our neophyte sun was growing it may or may not have been controlled by a decaying mass aggregate (similar to the present relationship between our earth and sun). The aggregate could have been growing as well, but that would only change the potential for separating one from another.

Because our neophyte sun has to utilize its gravity field to 100%, it will interact more and more with the dense mass zone it is approaching at the expense of interacting less and less with its aggregate controlling mass. Its speed toward this dense mass zone will increase, and as it penetrates and traverses this zone, its inertia–entropy characteristics will change. (The reason for such changes are intuitive based not only on variations of mass density, due to variations in size of heavenly bodies it passes by, but also that its media density in front of it will be more dense than the media it is leaving.) We need to explain why our sun is presently spinning:

One way our neophyte sun can do this is to suffer a reduction in linear kinetic energy for an increase in angular kinetic energy about its axis of propagation. One would expect a loss in velocity if this were to happen. Also, much like a feather or maple seed falling, if our sun is under a state of acceleration, a portion of the increase in acceleration will end up as angular acceleration.

We could also keep our velocity (or even our acceleration) constant and allow the inertia of our sun to be converted into entropy in the form of angular kinetic energy. The reader might think of it this way — as our sun traverses a more and more dense environment (at constant velocity or constant acceleration), more and more of its internal gravity (its mass – g_i^2) will be needed to ward off the stronger and stronger gravity field of its media. Much like our earth-moon problem, as this process continues the "packing fraction" of our neophyte sun will increase in direct relationship to the amount of gravity it has to ward off. In this way our sun will become more and more dense to adapt to its environment. During this process of varying g_i^2, our sun is expected to begin to rotate due to varying bulk shear modulus of these gravity interactions, inertia will be transformed into angular kinetic energy. The total entropy of our sun will increase at the expense of the inertia, while the remaining inertia will collapse into a *black hole*. Any of these scenarios will accomplish the needed spin.

At the photon level:

> Recently I was discussing a similar problem of a photon passing through a pane of glass – the photon apparently emerges with the same frequency and velocity but with less linear momentum. The photon need not give up energy to the glass, the glass may transform a portion of the photon's inertia into angular kinetic energy. This may be discernible if photons having passed through glass media have a greater resistance to refractive or diffractive bending. They may act as tiny gyros. They may also have varying reflective properties if spatial shear occurs between the photon and the reflecting mass – giving rise to a torque phenomenon.

Also, much like a planet being shot out the pole of a sun, our neophyte sun may have received its spin by being shot out of a pole of a larger spinning solar system. All these processes require a shear modulus between spatial gravity vectors to initiate the spin.

Now as our neophyte sun begins to leave this denser mass zone (i.e. it survived this transformation process, and its linear kinetic energy is now carrying it into a less dense mass zone), it will have differential inertia due to its spin. The greatest differential of its inertia-to-its-environment will occur at or near the pole leading into this less dense mass zone because the mass at the pôle has less motion. This differential depends to a certain degree on where adjacent heavenly bodies are arranged. The less dense the environment the better. This scenario differs from past theories because it does not rely on a large sun to pull the mass of the planets away from our neophyte sun. It is unique in that planets will be ejected out from the polar region of the sun due to our neophyte sun reaching a critical mass state. Much like the process of over-massed protons emitting photons as they leave a particle accelerator, or radioactive elements emitting particles, our neophyte sun will eject some of its mass (or diffuse its mass) to coexist with this new environment; furthermore, as our earth is being ejected to a high solar latitude orbit, our moon may be ejected to a high earthly orbit by a similar process. (Hudson Bay may be a geomorphological remnant of this? And may have provided a basin for the Archean and Algonkian rocks there. The last theory that I heard is that the Bay is an ancient meteor crater, and I haven't heard much else on it. The Bay's present volume, is *not* sufficient to generate the mass of the moon using 3.0 gr/cm^2. Also, I haven't kept up with the controversy of – was our moon ejected from the earth? Or captured by it? Is there any evidence that the Bay's basin-like geomorphology pre-dating the metasedimentary rocks there was of sufficient volume to generate the moon?)

Upon the sun's ejection of a portion of its already spinning plasmoid mass, the mass will either be shot out of our solar system or captured by it as a high latitude spinning satellite as our sun moves off into space. The angle of our sun's axis to our galactic disk's axis may vary from perpendicular, to tangential, to radial throughout the time span encompassing all planet ejections. (The axial tilts of our planets may depict these periods of ejection).

The actual ignition of our sun may take place during one of these ejection events or long after as natural radioactivity of its dense elements (caused by the sun's expansion) produce its hydrogen-helium atmosphere. It would appear that our earth is presently doing this – its density is decreasing causing radioactivity. At some time in the future our earth will grow large enough to retain all of the hydrogen and helium it produces by radioactivity. Then too, it will be a neophyte sun ready for satellite ejection, and/or ignition.

Note that this process is a conservative one in regards to redistribution of the total energy in our solar system. Our total

E_t being subordinate to those environmental processes which allow ejection and solar emission into our neophyte sun's exterior environment. Note too that at the roots of this intuitive format is life for it has reproduced our neophyte sun (our earth).

Presently our earth's relationship to the sun is: As our sun's surface moves to our right, our earth's northern hemisphere moves to our left. There is a constant shear of gravity net-cancellations as this process continues, and with that our earth is constantly being slowed down. This shearing action (drag) phenomenon would also explain why planets rotate faster with increase in radius from the sun.

This shearing action does not exist when a planet is shot out of the sun's pole – both masses will be rotating in the same direction if their axes of rotation remain parallel. As the planet begins to spiral down in latitude a decrease in rotation should be expected based on this drag which seems to be a direct function of the cosine of the solar latitude. If we place our earth back into its high solar latitude (keeping the distance from the sun to the planet the same as present) one would expect an increase in rotation due to both conservation of kinetic energy, and less gravity shear. Certainly many factors come into play with such a top-like paragenesis of our earth; How dense was this matter upon ejection? How much has it expanded to decrease its rotational velocity? The key to such questions may never be known, but closer examination of our solar system using this format may reveal more knowledge. The sequence of planet formation is important and probably developed quasi-radially outward leaving the faster and faster rotating planet younger and younger to the sun's equatorial plane.

3.3.4 GENESIS:

3.3.4.1 Introduction:

As to our ancient earth while it was in these high solar latitudes, it is here that I wish to intertwine this theory's postulations with the Biblical accounts found in Genesis. I warn you, I have been warned by both secular and clerical sages not to do this; however, I do believe the allegorical writings therein have some absolute truth, even though this truth may always contain a mystical parameter. So please read this portion with some sense of humor, and also with some sense of recognition that the physical statements made in Genesis do parallel what could happen during such a paragenesis. Because the Genesis account of evolution was with mankind long before Darwin regenerated another, I have chosen to use it as the framework to divide the earth's paragenesis. I did in fact wrestle with Genesis for some time to obtain this continuity because the Genesis account apparently says that the earth was made before the sun. The sun being made on the Fourth Day – I'll get to that then.

3.3.4.2 Day One:

Genesis 1:1 "In the beginning God created the heaven and earth." Depending on what meaning we put on Genesis 1:2 "And the earth was without form and void" will establish the sense of chronology in Genesis 1:1. Genesis 1:1 may be a statement of summation of all things having happened prior to the creation of the earth; or it may also be read as a summation of all things about to be discussed. "Form and Void" could mean that the universe, solar system, our earth was at that time a metaphysical concept within God, or these terms may mean physically irregular in shape and barren of seas, plant and animal life, etc... I have interpreted Genesis 1:1 to mean "In the beginning God created space and matter", or our universe, its E_t, its metaphysics, and all the motion therein. It would not seem unusual to start with the largest physical entity we can perceive. This is also consistent with the theory's subordination principle.

The other big event of the day was the creation of light. Because Genesis 1:4 reads "He separated the light from the darkness", I have interpreted this to mean physical-visible light; thus establishing the Jewish day – starting with a period of darkness ending after a period of light. The source of this light was our sun in a state of ignition.

This light could also mean some large sun from which our solar system paragenetically shot out from, but because if we look ahead to the Second Day we see the earth's atmosphere being formed, I chose not to telescope our sun's early paragenesis with such recurring events. This theory does not need a "big bang!" to initiate our existence; creation is sufficient.

It follows that after the earth was shot out from the sun's leading pole, the earth began to consolidate its constituents. With the constant deceleration of the sun's gravity field pulling at our south pole, the more dense materials of our earth ended up in the southern hemisphere or "gondwonaland" was formed. (This does not preclude some land masses being formed in the northern hemisphere; after all such a solar tide would cause some inertialization to the opposite side, or northern hemisphere of our earth.) As the earth indurated the lighter fractions H_2O, CO_2 etc. froze as an ice cap on the northern hemisphere. Note that with this arrangement a portion of the southern hemisphere would be in constant sunlight, some of it would be in a day–night environment and the northern hemisphere would mostly be in constant darkness. This is a physical interpretation to Genesis 1:4 "He separated the light from the darkness" the more obvious is that light and darkness are two separate phenomena. Physical evidence of this solar arrangement may rest in rock types after we reconstruct this gondwonaland; were its polar regions deserts?; were its coasts glaciated?

3.3.4.3 Day Two:

How would an atmosphere and seas be formed from this arrangement? As the earth began to spiral down in solar latitude, due to gravity drag and/or expansion. (This process directly links the orbital kinetic energy to the earth's rotational kinetic energy which will either decide or contribute to the shape of its solar sphere. Transfer of energy from one process to the other should be expected based on gravity shear.) When the earth descended in its solar latitude, its northern hemisphere was exposed to more and more daylight due to its slower rotation and descending solar angle. As this process continued our Second Day of Genesis was completed, providing an atmosphere of CO_2, H_2O, and those gases (fluids) which have low boiling or melting points, as the northern polar cap was melted.

3.3.4.4 Day Three:

The waters of the Third Day probably migrated to the equatorial portion of northern hemisphere Genesis 1:9 "unto one place". Genesis 1:9 goes on to say "let the dry land appear." This may mean that the northern continents were being exposed by the melting of the ice caps). One would expect much continental shifting as this mass transfer took place; as well as a loss of rotational velocity due to water transfer from the polar region to the equatorial region. The warmer latitudes (probably the latitudes south of the equator) would have been ideal for plant growth. Also, because there may have been no oxygen to support animal life, the earth may have had to incubate plant growth to provide this oxygen. There is a direct linkage between plant life – to coal – to graphite. Many major graphite deposits occur in pre-Cambrian rocks; however, most are not considered to be of plant origin. Descriptions of these pre-Cambrian graphite deposits by size alone resemble coal fields. Were the graphite zones of Madagascar and Ceylon derived from ancient coal fields? Graphite is less dense than most rocks, a lubricant and flakes or separates easily — large deposits could provide an excellent media for distortion and intrusion; furthermore its flake characteristics could allow any hydrothermal media to carry it much farther than ordinary country rock fragments in open conduits. Wadia (Ceylon Dept. Mineralogy, Records, Prof. 1, p. 19.) suggests Ceylon's graphite was "squirted or molded" into place. (Recently, I had the opportunity to analyze the paragenesis of the Big Four Mine, Kremmling, Co. There, one of the late stage events was that: hydrocarbon volatiles from the basement rock complex migrated up the still open conduits, coating all the sulfide minerals with a now sticky resin. Similarly, magmatic stoping of pre-Cambrian rocks containing disseminated graphite flakes and/or coal beds may have concentrated this graphite by "flotation" in melt residuals, to be later squeezed into open or opening conduits. When the flow regime in these open conduits was insufficient to transport these flakes out of the system, they would have been deposited as a magmatic phenomenon. If this scenario has any validity, this graphite's source would be better described as an organically-derived–magmatically-deposited event. Furthermore, magmatic stoping through a coal bed may explain why carbon-rich rocks are associated with some of these deposits. Could the coal macerals have de-oxygenated the melt to provide a stable environment for graphite flotation? Have we overlooked something here?)

The Third Day apparently brought about the plant kingdom, and while I recognize the need for plants to generate food and oxygen for the animal kingdom, I also recognize the lack of evidence of trees in pre-Cambrian time — algae, sponges, fungi, and foraminifera, yes — trees no. Genesis 1:12 "And the earth brought forth..." describes an evolutionary process not an instantaneous one. Was the Third Day, the advent of God's hand starting the processes that would later lead to Angiosperms in the Triassic period, or did these Angiosperms exist in pre-Cambrian time and have only been recognized as early as Triassic? In the former problem where the radius of the earth's orbit was decreased to r_o, and w_o increased to *3.33 w_1*, this corresponds to an 80% increase in the velocity of the earth's surface through its atmosphere. If this increase has any historical reality, I do not think tall trees were growing in pre-Cambrian time (save for non-desert polar regions). Wind erosion would be significantly more prominent. Genesis 1:12 "having seed each one according to its kind." The word **kind** need not mean absolutely identical seed, but may mean having-the-genes-of, or unto its parents; thus, one would not expect turnips to start growing pears without several minute changes in between, brought about by as many generations responding to their changing environment, i.e., exercising their metaphysical ability to survive by changing their inertia–entropy characteristics. I get a kick out of "theologians" who believe with only scriptural interpretation (not divine knowledge) that their all powerful God could **not** have created them by evolution. At any rate, I believe that the Third Day describes pre-Cambrian life. If gondwonaland is reconstructed, close attention should be paid to the latitudes of these now graphite deposits.

3.3.4.5 Day Four:

The Fourth Day was a period of disruption, it is also a stickler and seems to contradict the events postulated so far. (There was a reason why those sages told me to leave the Bible alone!)

I believe the verses of the Fourth Day say that the various heavenly bodies already in space were taking on their true significance as the earth was approaching the sun's equatorial plane. These heavenly bodies were for "signs and seasons" or life stimulators. The concept here is that signs and seasons do not come about without contrast, and contrast will come about only as our earth spirals toward the sun's equatorial plan; furthermore, suns, moons, and stars would not be visible to plants if the earth was blanketed with a thick atmosphere of "mist". The importance of this "mist" rests in Genesis 2:5 "for the Lord had not caused it to rain upon the earth", and Genesis 2:6 "the earth was watered by a mist". Is Genesis 2:5 past perfect? Or relative to Eden alone? Also, mankind did not document the rainbow until after the flood – perhaps when the sun's direct rays could penetrate the earth's atmosphere. How this atmosphere occurred without raining is not clear. Perhaps it simply formed from the polar ice cap by sublimation. The cyclical phenomenon of polar–ice–cap to flooded–ocean–basin has remained to recent times. In between these states, the extreme state of an atmosphere completely of mist which is translucent enough to sense the heavenly bodies may exist. Collapse of such an atmosphere would eventually bring about seas again and perhaps cessation of most plant growth by flooding, followed by desert formation. The "Garden of Eden" apparently developed under a "mist" environment which ended with Noah's flood. Note too, that had the moon been in a high latitude orbit above the earth it would rule over the darkness. I believe that the Fourth Day is the event completing pre-Cambrian time which brought about the Cambrian seas and the proper environment (oxygen and food supply) for animal growth.

3.3.4.6 Days Five and Six:

The Fifth Day included the Paleozoic era, and the Sixth Day gets us to mankind. Such an interpretation does not leave too many glitches between geological evidence and the Bible's Genesis.

Whether the Fourth Day was an event or consumed a portion of late pre-Cambrian time or early Paleozoic time is not clear because no flora and fauna species are mentioned. Geologically the Fifth Day may be terminated by the Carboniferous period because some land animals started showing up in the Permian period. The whales and fowls mentioned in the Fifth Day may be large fishes and pterosaurs. Divisions are somewhat arbitrary depending on interpretations of scripture and future fossil exposure; however, the Bible's general trend of plant, to sea life, to more advanced sea life, to land animals and to more advanced land animals is remarkably similar to our geologic chronology of the same evolution. The author feels that past geologists have depicted the types of pre-Cambrian flora fairly correctly; however, the total amount of organic material deposited during pre-Cambrian time is seriously questioned. Ponder this: how did the author of Genesis 1:25 know that cattle were one of the last species to come to earth.

During this time (second day to present) gondwonaland was probably breaking up. As the shearing action between the gravity fields of the heavy portions of the earth's crust and the sun increase due to decrease in solar latitude, gondwonaland began to separate into continents which migrated northward toward the equator. (A transformation of entropy forms – loss of rotational speed at the expense of a simultaneous increase in rotational kinetic energy by the continents moving to a higher state of kinetic energy at the equator.) Glaciation could have taken place during this time either by these plates under thrusting polar ice caps or pseudo glaciation as the result of unconsolidated rock masses slipping over shield rocks without the vehicle of ice being present.

The need for computer modeling becomes ever present as each fact needs to be located and chronologically sequenced with the evidence presented by remnant magnetism, continental drift, glaciation, life forms etc... Note: If remnant magnetism is a function of the earth's magnetic field being perpendicular to the sun's ray, at high solar latitudes the earth's magnetic pole will be much lower in latitude. It was once near Hawaii.

Because some of the change in rotational speed can be attributed to increased gravity shear with lower solar latitude, the original surface area of the earth was probably larger than 30% of the earth's present surface.

3.3.5 OTHER QUESTIONS

Other questions and ideas that arise from this type of earth paragenesis are:

3.3.5.1 Sun Spiral

Astronomers have indicated that the sun is spiralling through space, and it probably is; but if our earth is in a latitude spiral will that alter our perception of our sun's spiral?

3.3.5.2 Similar pre-Cambrian Rock Types

If gondwonaland represented a land cap on the southern hemisphere, sedimentation along the coast of this land mass might be quite similar with facies changes not very distinguishable. Its ocean perimeter would represent a quasi-constant latitude.

3.3.5.3 Ocean to Land H_2O Transfer

One might not expect a lot of ocean-to-land transfer of water as we know it today. The wind direction would be quasi-parallel to the coast of this land cap.

3.3.5.4 Man's Creation

A theological premise was found while studying the book of Genesis using only chronology to determine the paragenetic sequences of evolution. Adam and Eve were probably not created on the Sixth Day of Genesis; rather, God probably created them after the Seventh Day. The race of man, male and female, created on the Sixth Day would appear to be one created as a product of God's evolution (they were animals without man's higher consciousness of God). The reason is this: God rested on the Seventh Day – a period of inactivity – but since God created Adam and Eve, He has always been in a state of activity (sometimes not the kind He wanted) – scripture, post Adam and Eve, speaks of no period when God rested; thus it would appear that the "Sons of God" spoke of in Genesis 6:2 were really Adam's sons who married female species of evolved man. Genesis 6:4 indicates this "commingled – being of soul and flesh" shall only exist on the earth for a finite period of time (120 years). This prophecy may have two time tables involved: 1. One hundred twenty years later (earth time) the flood destroyed the race of carnal man; 2. One hundred twenty years later (by another time level), Noah's seed will vanish. Because Noah's lineage was the only one to survive the flood we are all a commingled product of Adam.

Adam probably was an evolved man whom God breathed a soul into (God baptized Adam with a higher order consciousness of God); after all, he created Adam from the ground the same place as he created all the other species. Theologically, there seems no conflict with this scenario, though it may at first ruffle a few Biblical pages! God being omnipotent has the power to create man by evolution, and also, divine providence. Perhaps it is fitting that we be a commingled product of both, sharing commonness with the other species, and commonness with God. Without the baptism of God which directs our energies toward Him (no other animal does this), we might have destroyed the earth long ago. Although this scenario offers more continuity between evolution and mankind, it must be said that it didn't come to me by conscious impartation of divine knowledge; besides, I find theology best when my spirit is free to be enveloped by the mysteries of religion – perhaps that's why I'm Catholic?

3.3.5.5 Atomic Structure

Using spatial spheres or shells may be a better way to model atoms. I find the present means of describing them by protons with paired electrons a bit baffling. The reasons are:

Take a sodium ion – Na^{+1} – if we try to pair its eleventh proton by putting another electron on to it – the electron ends up in a higher orbital and the atom's electron complex increases in size. I assume, and perhaps wrongly, that this neutralizing electron ends up farther from its paired proton. Like gravity pairs, the centripetal force to keep this last electron attached to the atom is a function of the inverse square of the distance separating it from its paired proton in the nucleus. So why is the electron more stable farther away from its pair? If we say that the lower shells have sufficient negative charge saturation to repel this electron into a higher orbit, then these lower shells also have enough repelling energy to kick the electron out into space; however, if we construct our atom from a mono-substance, inertialized gravity (mass) and active gravity (field energy), this problem becomes much more simple. The mass saturation in the lower orbit will prevent this electron from penetrating this region and repel this electron to a higher orbit. (The atom diffuses its mass.) Now the electron has both the nucleus and mass of the 10 electrons to orbit about; thus it can orbit at a higher state of kinetic energy. Once again – are the nuclei of our atoms in part *black holes*, and as such, does nucleation stem from the aggregate impingement of gravity net-cancellations about its centroid? I think yes.

3.3.5.6 Speed of Light

A special problem recognized in this paper is: Why can light only travel so fast? I chose to explain this phenomenon by bulk shear interactions of gravity fields. In terms of Einstein's relativistic physics, this phenomenon may be accountable by our relative movement through space. I don't know if Einstein or anyone else has sufficiently explained this phenomenon per the Theory of Relativity.

3.3.5.7 de Broglie's Equation

I believe there is much more to de Broglie's equation than appears. The rapidity of an object's oscillatory existence will determine the compatibility of it to share the same space of another object, and therein the bulk density (bulk gravity) of their aggregate will vary as viewed from a transcended frame.

Example:

When a fan is stationary, you can place your hands between its blades, but if the fan is turned on, the fan will not allow your hand to be placed between its blades. (This may seem dumb but it shows how oscillatory variations may effect the bulk density of any given region of space.) When various oscillations are present within a suite of objects certain communions of the corporeal/meta-corporeal will not take place naturally.

3.3.5.8 Show and Tell

1. If gravity can implode toward the center of the earth, and also reach out ethereally to you, the moon, and the sun, do you see why about 50% of its substance must be bound internally, and 50% of its substance be free to externally interact with the earth's environment? Do you see why I postulate θ will hover about **45°** under normal conditions?

2. In nuclear β^- decay a neutron divides into a proton (which remains in the nucleus), and an electron (which is shot out of

the nucleus). The difference in mass of the neutron and resulting proton is $2m_e$ (m_e = **mass of the electron**). Do you see why the **Unitive** equation predicts this?

Explanation:

The principle of equivalence states that when you are in free fall or free orbit you do not feel the force of gravity, i. e., you feel weightless when falling. Consider that gravity is a substance and not just a naked force. Consider further that your gravity field is exterior to you and in **principle of equivalence** with the earth's, i. e., you do experience the substance of your gravity field and the earth's as inertia (mass to the sun), but you equate this experience as a naked force binding your Newtonian mass and the earth's together. You might say that you cannot determine your exterior gravity substance while you are using it to determine your Newtonian mass. Similarly you cannot determine your entropy in the **Unitive** equation due to **principle of equivalence**, i. e., you do not sense the kinetic energy you possess due to your rotation about the earth's axis, and your orbital velocity about the sun, etc...

When the electron is part of the neutron in the nucleus, and you pick up the entire atom, you pick up both the neutron's Newtonian mass, and its gravity field, i. e., at $\theta = 45°$ in the neutron, in the nucleus, 50% of the electron will be in Newtonian mass and 50% of the electron will be in its gravity field; however, both will appear as Newtonian mass to you. When you intercept the electron (by itself) after it is emitted from the atom, you will do so at the **principle of equivalence**, and you will only perceive the electron's Newtonian mass. I postulate that the neutrino that physicists have been searching for is the electron's gravity field expanding or coming into equilibrium with its environment.

Note: Because it has been shown that the **gravity** event is a **charge** equivalence, the neutrino can have a spin function.

The energy given to the environment by the electron, should follow a $cos^2\theta$ curve, with the missing heat energy in **the Ellis and Wooster experiment** being absorbed by the environment to overcome the environment's latent heat or inertia - much like the photoelectric effect, or when you pick up a ball, the binding inertia must be overcome before you will experience heat or kinetic energy.

3. Given: The old relativistic equivalence,

$$E^2 = (mc^2)^2 + (pc)^2,$$

which is a combination of the

relativistic momentum, $p = \dfrac{mu}{(1 - \frac{u^2}{c^2})^{1/2}}$, and the

relativistic kinetic energy = $\gamma mc^2 - mc^2$,

show that E^2 above can be easily obtained from the **unitive** equation,

$$E_t = mc^2 \cos^2\theta + hv\sin^2\theta \ @ \ \theta = 45° \ \text{for} \ E_t.$$

Hints: 1. Use deBroglie's equation for h and
2. Maintain E_t's external relativity @ $\theta = 45°$ by maintaining θ's internal **unitive** directions.

4. On Page 61 it was shown that $G = 4\pi a_0$ when unitized in the cgs system; however this is not true in the MKS system, or any system that is a cgs system multiplied by a power of 10^n. I point this out here because this inequivalence existed long before the **unitive** equation.

(Excluding units)

The MKS system unitized value for $G/4\pi = 5.30 \times 10^{-12}$, or a power of 10^{-1} more negative than the MKS system Bohr radius $a_0 = 5.29 \times 10^{-11}$.

At first I thought this was an inequivalence due to the asymmetrical expansion of the cgs system to the MKS system, i.e., the MKS system is $cm \times 10^2/gr \times 10^3$, so I symmetrically expanded the cgs system value for $a_0 = G/4\pi$ and compared it to the expanded Bohr radius value below in **Table 3–3**.

TABLE 3–3:

$a_0 = G/4\pi$ unitized to n		$a_0 = a_0 \times 10^{-n}$	
n = 0	5.30 x 10⁻⁹	n = 0	5.29 x 10⁻⁹
n = 1	5.30 x 10⁻¹¹	n = 1	5.29 x 10⁻¹⁰
n = 2	5.30 x 10⁻¹³	n = 2	5.29 x 10⁻¹¹
n = 3	5.30 x 10⁻¹⁵	n = 3	5.29 x 10⁻¹²

Table 3–3: Comparison of a_0

We see then that a_0 derived from G behaves as a quantum function as it should; therefore I postulate that the unitized Bohr radius changes with the system used by:

$$\frac{(R_{cgs})(10^{2n})}{(R_{cgs})(10^{3n})} \times a_{0_{cgs}} \times 10^{-n}, \text{ where } n = 0, 1, 2, 3, \text{ etc...,}$$

and that in the asymmetrical MKS system the unitized Bohr radius = 5.29×10^{-12} meters.

a. Show that $(R_{cgs})(10^{2n})/(R_{cgs})(10^{3n})$ is the cgs ratio of surface area to its cgs root, i. e., $\frac{4\pi R^2}{4/3 \pi R^3} : \frac{4\pi (R10^{2n})}{4/3 \pi (R10^{3n})}$.

b. Given: that the Bohr radius changes (in part) as a ratio of surface area / volume — could this explain why the Bohr model broke down as the model expanded from hydrogen?

c. Given: the force of charge for the hydrogen atom unitized to its own Bohr radius,

$$F = \frac{k_0 Q_1 Q_2}{R^2} = \frac{k_0 e^2}{M^2} = \frac{(8.98755 \times 10^9 \, kgM^3/sec^2 - C^2)(1.60210 \times 10^{-19} C)^2}{M^2},$$

$$F = 2.306856 \times 10^{-28} N.$$

Show that this value can be derived from the **inverse** relationship between gravity and charge shown on **page 61**:

$$e^2 = \frac{4h^2 \in_0}{Gm_e} \text{ where } k_0 = 1/4\pi\in_0, \text{ and } G = 4\pi a_0 = 6.65 \times 10^{-10}M,$$

$$F = \frac{k_0 e^2}{M^2} = \frac{(1/4\pi \in_0)(4h^2 \in_0)}{M^2(Gm_e)} = \frac{h^2}{G\pi m_e} = \frac{h^2}{4\pi^2 a_0 m_e},$$

$$F = \frac{(6.6256 \times 10^{-34} NMSec)^2}{4\pi^2(5.29167 \times 10^{-11}M)(9.1091 \times 10^{-31}kg)M^2} = 2.306867 \times 10^{-28}N;$$

thus we no longer need **charge** to solve "electrostatic" problems.

3.3.5.9 Partially Solved Problems

1. Considering that most equations for energy, E, involve a physical constant multiplied by a spin function (i.e., spin function meaning a resonant relationship between distance and time – ft²/sec², cycles/sec, etc...), I had visions of finding such an equation for, or within, G, the universal gravitational constant.

One day while "tweaking" my unitized equation, $G = 4\pi a_0$, I thought to myself – "maybe there is an equation within this equation"; So, I took the square root of a_0:

$$(a_0)^{1/2} = (G/4\pi)^{1/2} = 7.29 \times 10^{-5} \text{ (units-?)}.$$

Not recognizing this number, I went to my Halliday & Resnick physics book, and I attempted to open it to its table of Fundamental and Derived Constants, Appendix A, p.31. A strange thing happened; my book didn't open to this page as it usually does; it fell open to Appendix A, p.33, and there was this number:

7.29×10^{-5} (?).

Do you know what units the question mark represents? It is the spin function of the earth in radians/sec, or

$$a_0 = \left(\frac{2\pi}{T}\right)^2, \text{ and } G = 4\pi\left(\frac{2\pi}{T}\right)^2 = \frac{16\pi^3}{T^2}.$$

Even though I am not disciplined in quantum mechanics mathematics, I attempted to use this equation to show that each of our planets is quantumized to its orbit by a quantum variation in this equation. It works! Unlike the Lyman-Balmer-Paschen series for hydrogen, the solar system can be divided into two quantum series which results in a constant. It is interesting to note that the series changes between Mars and Jupiter where the asteriod belt exists. Also, as shown in **Table 3–2**, page 58, there is a large increase in "relative charge" between Mars and Jupiter.

I simply modified the above equation to take on the relativity of each planet:

$$G_p = 4\pi \left(\frac{2\pi R_0}{T_0}\right)^2,$$

where G_p – represents a gravity event,
R_0 – radius of each planet's orbit (in centimeters), and
T_0 – time of orbit (in seconds).

The first series is represented by:

$$G_p = n\, 4\pi \left(\frac{2\pi R_0}{T_0}\right)^2, n = 1 \rightarrow 4, \text{ for Mercury} \rightarrow \text{Mars, respectively.}$$

This series will also accommodate planets, Jupiter→Pluto, if **n** is allowed to take quantum leaps, **n = 13, 26, 52, 78, 104**, respectively. If we do this, we will end up with a little tighter system, i.e., below in **Table 3-4**, $\bar{x} = 2.9834$, s = 0.1533, instead of the $\bar{x} = 2.9292$, s = 0.1801 listed. Note that I allowed these quantum leaps to be multiples of thirteen (13), whereas **n = 13, 25, 50, 78**, and **103** give even tighter results; however, at this time I feel more theory can be gained by using a constant multiple which brings us to the second series below:

The second series is represented by:

$$G_p = 2n\,4^2\pi^2 \left(\frac{2\pi R_0}{T_0}\right)^2, \ n = 1/2, 1, 2, 3, 4, \text{ for Jupiter} \rightarrow \text{Pluto},$$

respectively.

This second series is implaced here for theoretical reasons; even though it does not fit the data as tight as the first series, it is postulated that because these major planets consume some of their gravity for local satellite activity, that G_p will decrease, and we may be looking at a gravity sphere, within a gravity sphere; thus, our quantum number should be a multiple of 4π, i.e., a multiple of the surface area of these spheres with unitized radii as in our beginning equation, $G = 4\pi a_0$. See **Table 3–4** for results.

TABLE 3-4:

Planet	n	R_0 (cent.)	T_0 (sec.)	$G_p = 4\pi\left(\frac{2\pi R_0}{T_0}\right)^2$	G_p	$\bar{x}$	s
Mercury	1	58 x 10^{11}	7.6006 x 10^6	$G_p = n4\pi\left(\frac{2\pi R_0}{T_0}\right)^2$	2.8889 x 10^{14}		
Venus	2	108 x 10^{11}	19.414 x 10^6		3.0706 x 10^{14}		
Earth	3	149 x 10^{11}	31.558 x 10^6		3.3177 x 10^{14}		
Mars	4	228 x 10^{11}	59.355 x 10^6		2.9281 x 10^{14}	3.0513	0.1939
Jupiter	1/2	778 x 10^{11}	374.34 x 10^6	$G_p = 2n4^2\pi^2\left(\frac{2\pi R_0}{T_0}\right)^2$	2.6929 x 10^{14}		
Saturn	1	1,426 x 10^{11}	929.59 x 10^6		2.9340 x 10^{14}		
Uranus	2	2,869 x 10^{11}	2,651.3 x 10^6		2.9200 x 10^{14}		
Neptune	3	4,495 x 10^{11}	5,200.2 x 10^6		2.7948 x 10^{14}		
Pluto	4	5,900 x 10^{11}	7,852.5 x 10^6		2.8155 x 10^{14}	2.8314	0.0989
						2.9292	0.1801

Table 3–4: G_p quantumized to a constant.

Note: that as in our "relative charge" calculations in **Table 3-2**, page 58, the central portion of both of these series have larger values than either the inner or outer portions.

Given the fact that a number of these planets have local satellites to consume their gravity as they orbit about the Sun, I find this quantumization to a constant G_p support of a unitized solar system complex. (Perhaps someone can use this quantum approach to show that the local satellites about the larger planets are also unitized.)

Given the old circular planet equation,

$$GM_s = \frac{4\pi^2 R_0^3}{T_0^2}, \text{ or } \frac{GM_s}{R_0} = \left(\frac{2\pi R_0}{T_0}\right)^2, \text{ upon substitution,}$$

$$G_p = n\, 4\pi \left(\frac{2\pi R_0}{T_0}\right)^2 = n\, 4\pi \left(\frac{GM_s}{R_0}\right),$$

$$G_p/G = n\, 4\pi \left(\frac{M_s}{R_0}\right),$$

where M_s – mass of the Sun (in grams), and
R_0 – radius of each planet's orbit (in centimeters).

This problem of unitizing the solar system is only partially solved here because, except for the Earth, the axial spin function for each planet has not been formulated as a quantum set within the original equation, $G = 4\pi a_0$. Perhaps someone better at quantum mechanics than me can solve this problem, as well as, tie in the local unitization of planetary satellites to show how they affect a planet's G_p.

As a side note: The equation, $G = 4\pi a_0$, indicates that when a hydrogen atom is subjected to a higher outside gravity event, its Bohr radius will increase. Picture a deuterium atom being brought to the Sun or to the surface of a dense metal; as the atom's environmental gravity event increases there will be a decrease in spin function between the deuterium nucleus and its electron, as they both give up "charge" to the environment. The shielding effect of the electron will decrease, and the probability of two like nuclei fusing will increase. In a general sense this may be the mechanism for the Fleischmann–Pons cold fusion process.

On a chemical level it may also be the reason why platinum makes a good catalyst for certain reactions, i.e., as atoms are forced into a higher gravity event, their electron clouds will be so warped that chemical reaction will take place with little or no heat energy to activate the process.

2. Perhaps it is not peculiar while writing an evolutionary work as this one to have a quirk keep reappearing. Indeed it has

happened enough times that it has almost become a personal friend. The quirk is simply this – my equivalences are usually off by roughly 1.003 to 1.0035.

In my g_{lat}. problem, page 29, my $G = 6.6506 \times 10^{-11}$ is low by 1/1.003, on the other hand $B = 0.01443$ is 1.003 larger than Planck's second radiation constant. My Bohr radius, page 61, $a_o = G/4\pi$ is 1.003 too large, and on and on, it goes.

It has long been thought that space is fairly isotropic. If we couple this thought with the unitive concept that gravity is the energy of all resonances from photons to galaxies, we might conclude that the universal gravitational constant, G, would be applicable throughout space. However when we quantify energy for photons we use Planck's constant, h, a number numerically similar to, G, but slightly different. 6.6256 vs. 6.670 (ignoring the powers of ten).

If we calculate $G = 4\pi a_o$ at the atomic level by substituting the real Bohr radius value of 5.29167×10^{-9} cm,

$G = 6.6497 \times 10^{-8}$.

This value, numerically, sits very close between h and G above, and incidentaly all three numbers are numerically 1.003 to 1.0035 apart.

Is G at the planetary level 6.670, at the atomic level 6.65 and at the next stable resonance level (the photon) 6.63? Is there a gradient to space such that large objects subordinate the small, i.e., is there a basis here for the law of subordination?

Suppose we envisage a photon travelling through an atom's electron cloud while the atom is in the Earth's atmosphere. What gravity event might we assess to this event? If we equate Planck's constant, h as unity, then:

G at the photon level = 6.6256×10^{-27},
G at the atomic level = $6.6256 \times 1.0036 \times 10^{19}$, relative to h,
G at the planetary level = $6.6256 \times 1.0067 \times 10^{19}$, relative to h.

If you will indulge me?

93

Multiplying all three above, and adding a power of ten gives:

$X = (6.6256 \times 10^{-27}) (6.6256 \times 1.0036 \times 10^{19})$
$(6.6256 \times 1.0067 \times 10^{19}) \times 10,$

$X = 2.9477 \times 10^{14} = G_p$ from the above planetary problem.

Could someone tell me why?

Oh, and incidentally our quirk is back, but a little larger.

$$\frac{2.9476 \times 10^{14}}{2.9292 \times 10^{14}} = 1.0063 ? \text{ Our (quirk)}^2.$$

Good hunting.

3.4. Quantifying Life

3.4.1. MATHEMATIC PREMISE:

Given: the **law of conservation of energy** which states that the amount of energy in the universe is fixed.

Let $E_t = (1)^a \cdot (1)^b \cdot (1)^\theta$.

Premise:

1. Internally an object is in a state of unity with itself; and

2. Externally an object is in a state of relativity with other objects.

It follows that the exponents a, b, and θ must address both (1.) and (2.) above simultaneously; therefore, these exponents are **arithmetic summations** of units comprising E_t, or (a, b, and θ are not power exponents).

It follows: $log_{e_t} E_t = log_1 (1)^a \cdot log_1 (1)^b \cdot log_1 (1)^\theta$,

$log_{e_t} E_t = a(1) \cdot b(1) \cdot \theta(1) = a \cdot b \cdot \theta.$

Example:

Suppose there are six bulls in farmer Jones' field. It is winter; there is little food; and the watering pond is froze over. Also each of the six bulls weigh the same. Jones comes to the far end of the field and drops off hay and water. Each of the six bulls begin to walk at the same velocity toward Jones.

Classically we might say that we have six unit-bull-vectors in the field, i.e., six scalar unit masses times their respective velocity vector.

Three months later we look out in Jones' field, and we see three pairs of bulls butting head-to-head (motionless). (For now we won't ask why the bulls are doing this observable fact?)

If we were to examine this problem using classical vector addition, we would conclude that there are six scalar bulls, but zero unit-bull-vectors in Jones' field, or:

$$3(\overrightarrow{bull} + \overleftarrow{bull}) = 3\ (0) = 0.$$

(In **unitive** thought there are still six unit-bull-vectors in Jones' field, or we could say there are three objects each comprised of two unit-bull-vectors in states of net-cancellation in the field. It is deduced that the inertia caused by these net-cancellations of three binary unit-bull-vectors is of physical consequence and not zero.)

If we were to use the scalar product we would have:

$$3[(\overrightarrow{bull}) \cdot (\overleftarrow{bull})] = 3\ |\overrightarrow{bull}|^2 = 3(1)^2 = 3\ \textit{bull-vectors}.$$

Somehow our six unit-bull-vectors became three (unit?)-bull-vectors.

If we were to say that we have three scalar bulls pushing in one direction times three scalar bulls pushing the opposite direction we would have:

$(3\overrightarrow{bull}) \cdot (3\overleftarrow{bull}) = 9$ *bull-vectors*.

Neither the vector addition or the scalar products above conserves; the amount of unit-bull-vectors in Jones' field. We have to intuitively go outside the mathematics to know there are six unit-bull-vectors in Jones' field.

Let us return to the first scalar product above:

$3[(\overrightarrow{bull}) \cdot (\overleftarrow{bull})] = 3 \; |\overrightarrow{bull}|^2$.

If we allow the exponent to be an **arithmetic summation** of the unit-bull-vectors involved in each binary couple, the problem is solved.

We would read:
Three units comprised of the sum of two unit-bull-vectors which equals six unit-bull-vectors.

We have conserved Jones' unit-bull-vectors. It is assumed that at the bulls level of freedom we cannot divide a bull into smaller bulls in states of net-cancellation, and the product above-means to be in a state of net-cancellation.

Note: This **unitive** mathematics solves another problem at hand. If we were to butcher one of Jones' bulls and eat a portion of it, we would receive free energy from its now scalar mass. In classical physics energy is a scalar, so classically by eating a portion of the bull we receive scalar energy from scalar mass; however, in **unitive** thought mass and energy are vectors because it is not sufficient to say we have a certain quantity of mass or energy, we must apply mass and energy to solve physics problems, and that application gives these classical scalars destiny or vectorship. In **unitive** thought we have the vector-mass of the bull being transformed to our vector-energy.

Note: Even though this mathematics conserves Jones' bulls, it tells you nothing of how much energy is being transformed into inertia by the bulls. Once again we would have to metaphysically experience (enter into) the problem to determine this inertia. This is true because lack-of-motion provides **n** roots of zero velocity.

It follows that the $log_{e_t} E_t$ must satisfy both the internal unitive condition, and the external relative condition above; therefore, $log_1 (1)^a$ can be said to equal zero (0) internally, but externally the $log_1 (1)^a = a (1) = a$.

(Though I refer to the above mathematics as "**unitive**," I feel that Einstein or one of his proteges must have previously developed it, otherwise the tensors they used would not have conserved mass and energy. If that be the case, I reintroduce it here because it is needed for the **Unitive Theory's** minimum set.)

1. Given: The unitive product of $E_t = (1)^a \cdot (1)^b \cdot (1)^\theta$ and $log_{e_t} E_t = a \cdot b \cdot \theta$;

Show that if the base $e_t = 1$ *kilogram (meter²/sec²)e²*, and your relative mass is 75 kilograms, *a* above has a relative value of 75 and an absolute value of 150.

2. Do you see why E_t is internally unitive and externally quantitative?

3. Given: That you are an $E_t = (1)^a \cdot (1)^b \cdot (1)^\theta$;
Write a logarithmic equation quantifying yourself such that you can be expressed as:

you = 75 · b · θ

Tell me: are you self-rightous if your θ value exceeds 1?

4. If $(1)^a = e^0$ such that the **natural logarithm** is $a \ln (1) = 0 \ln e = 0$, convince me that I cannot naturally reduce you to nothing!

Humble your θ.

4.0 CLOSING

I don't know if I will work in industry any longer; so I thought it best I get this on paper before the processes of life prevent me doing so. A physicist I met said to me "Science has invested so much time into the Theory of Relativity it will never change." Has Einstein's truth turned to demagoguery? This paper has been purposely disruptive for such reasons. My equivalence is Einstein's, only reshaped to preserve Thomas Aquinas' metaphysical parameters in life: God's demagoguery. I do believe that with the union of physical and metaphysical parameters we may find new truths. This did not come to me as a philosophical truth: For we cannot observe relative behavior, without simultaneously observing absolute behavior as well.

5.0 THE FOUR SEASONS:

[Epitaphs of Spring

(Conversations:

On a knoll in a western plain
The wind wained a song to fill
My ears with words of pain
It blew: Boothill! Boothill!

Stoic epitaphs staked each's claim;
Page by page, they insured this thief –
Time said "harrow this plain."
"Their cryptic treasures per your belief."

Did I hear Sir Isaac's levity call
To Albert, Max and the rest plainly
"Pompously he strides on us all
He knows: once we presumed as vainly."

"We have him now, and well,
For on this Western plane so spacious
His rogue-youth decreed our seed 'to hell'
How smug! How delightfully ostentatious!"?)

Initiation]

. . .

[Faith vs Summer In The Citadels

    (Arise! Come down!

        To your flock feeding contentedly
        At the slope where they've been led.
        They grow fat, sluggish, crying lamentably
        To one another. Your masochistic quadruped
        Governs! The spatial differentiation
        Of the non-radiating who have an affinity
        To radiate. Pressers of your emetic radiation,
        they tax your fraternal–trysts' serenity.

        Arise ! Come down !
        You students of Descartes planes
        Where "black-holes" transcend your math;
        Your centrifugal forces. Inertially arcane,
        Essential, creations of entropic rath!)

                        Breeds Contempt's Insation]

            + + +

[Autumn's Hope

(Youth:

I hope this finds you in a meadow
With my painted leaves scattered about you,
With the bird's vacating autumn calls,
As geese fly south for the winter.

(Perhaps it is in knowing:
You are free to fly
With life's ganders
That makes me kind of chilly.)

And if this finds you in the city:
Take to the meadow, wood and wind,
Watch these red, dead pages of life
Fall hopefully to their grave;

Be not disturbed however, my inchoate
Imbroglio will survive *Winter's Love*
And *Epitaphs of Springs* to come
Initiate its final destiny.)

Brings Moderation]

. . .

[Winter's Love

(Father:

>Chide not in my vain picture done,
>Seeking reverence in my matter's source
>This manly form of your Son
>Upon a painted cross I forced.
>
>A painted cross my hands might bare
>Your Son, they've not the Spirit to conceive;
>So this manly forum I placed there
>For seekers who saw not, yet still believe:
>
>Their Golgotha, for man's and my sin,
>Was to bare a cross, their nails within.
>I found Him there, dying to appease us.
>Rex dv Reges, hicest Iesvs.)

Unites Humiliation]

YOUR UNITIVE EQUATION

6.0 ACKNOWLEDGEMENTS

Jerome Benson: Physics; Duluth Central High School, Duluth, Minn. for stimulating an adolescent mind with abstruse contemplations. Oh! dear, dear and another, _____? Look at what your lab's created;

Mike Sydor: Physics; U.M.D, Duluth, Minn. for your sense of humor during those confrontations, may every drift you make be full of silver;

F. B. Moore: Head of Chemistry; U.M.D., Duluth, Minn., advisor, for sound advice when I wouldn't listen. I saved the verse to last this time;

Bill Van Matre: Head of Mining; Montana Tech., Butte, Mont., advisor, for your inordinate capacity to hustle the best for your students, and the best out of them;

Fred Earll: Head of Geology; Montana Tech., Butte, Mont., advisor, for loaning me more than money, may its interest be yours;

Hugh Dresser: Assoc. Prof. Geology Dept.; Montana Tech., Butte, Mont., for teaching me the dynamics of three dimensional – paragenetic events. I have to single you out as my best teacher ever, and to date, the finest geologist I know. May those neon signs never cease flashing;

Mike Doman: Physics, Montana Tech., Butte, Mont. for restimulating in me what Mr. Benson started. Although I never went into physics, perhaps you will be glad to know it's never left me alone;

John McGuire: Humanities, Montana Tech., Butte, Mont. to the degree this vain treatise conveys faith, hope and love to this and future generations, your writing classes deserve much of the credit; however, as to its lack of discipline, I offer you a field day, but please, spare me your bloody copy;

Sharon Nolan-Rosager: Alumni, Montana Tech., Butte, Mont. if John has gotten this far, internally, my equation views it as a success. Thank you for being the mother of my work; and

Chuck Herdon: Assoc. Prof., Engineering Science, Montana Tech., Butte, Mont. To the day when I was meditating on the steps of Main; you rushed by saying "Behling watch-ya all doing"; "Trying to figure out what gravity is:" I said; "Forget it." you replied, "there's better things to do." You were right, but thanks for its stimulation.

7.0 EPILOG

In 1978 I tried to reintroduce God to science; seven years later, I formed an equation in His image.

We have the internal freedom to meet God in the Spirit, but we cannot have our neighbor show Him to us, rather our collectivism gives us what might be glances of His presence through faith, hope and love fellowships.

Fellowships also bring controversy, and in that light there were those vain times when I felt like Moses seeking a Pharaoh's throne, and even those more vain times when this vessalage was Pharaoh indulgently feeding on the discord of Moses' long past, and so many times I starved for clues they left as to who I was; and why I should be engulfed in abstruse contemplations; and even though I have left you enough controversy to shout thou blasphemer; thou fool; there was that transcendent day when each time I closed my eyes another seeker from the past came and shook my hand. Why? I do not know.

I only know in faith, hope and love, I am rid of it.

ADDENDUM

Dual Function Equation

By

Thomas A. Behling
July 20, 1978

Copyright 1978

CONCEPT

The author chooses to begin his concept with the postulation that "pure mass" is a net-cancellation of "pure energy" such that for "pure mass" vectors to occur, "pure energy" vectors must decay according to $dM_{pure} = -dE_{pure} / c^2$ as Einstein predicted. "Pure mass" is considered a vector state in which no field energy exists, and "pure energy" is considered a vector state in which no mass exists. Neither "pure" state should be confused with common matter or common field energy systems as man perceives them in everyday occurrences; i.e., matter and field energies are resonant systems of "pure mass" and "pure energy" vectors dynamically oscillating from one "pure" state to the other. Based upon using "pure mass" and "pure energy" vectors as discussion tools, this paper will be concerned with the topics on the following page:

Contents

I. DERIVATION OF THE DUAL FUNCTION EQUATION FOR PHOTON TRANSFER 1

II. DERIVATION OF PHOTON MOMENTUM 3

III. DISCUSSION OF REFRACTION AND DIFFRACTION PHENOMENA 7

IV. DISCUSSION OF MEDIA VS NON–MEDIA TRANSFER OF PHOTONS 8

V. DISCUSSION OF REFLECTION PHENOMENA 10

VI. DISCUSSION OF ENTROPY AND INERTIA 12

    Introduction 12
    Discussion 13
    Summary 17
    Discussion on Disorder 17
    Further Discussion on Disorder 19
    The Need for Motion 21
    Variation in the Mass Function Due to Motion 21
    The Velocity of Light 24
    Gravity 26
    Second Law of Thermodynamics 27

VII. DISCUSSION OF THE *THEORY OF RELATIVITY* 28

VIII. CONCLUSION 28

I. DERIVATION OF THE DUAL FUNCTION EQUATION FOR PHOTON TRANSFER

A. Let $\vec{E}_t$ = the total energy of a photon such that $\vec{E}_t = \vec{M}_t c^2$ where $\vec{M}_t$ = the rest mass of a photon $\neq 0$.

B. Let $\vec{E}_t$ be a conservative vector revolving about the origin in a cyclical mode as shown in *Figure 1*.

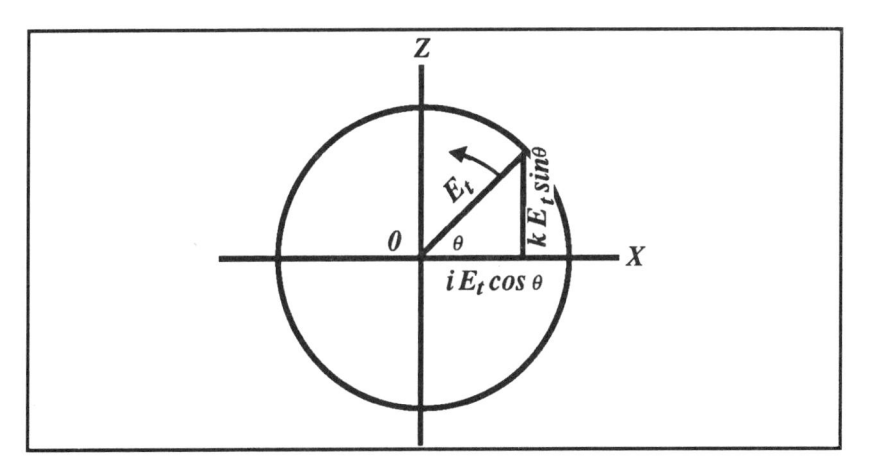

Figure 1.

Figure 1, $\vec{E}_t$ *a conservative vector revolving about the origin in a cyclical mode.*

It follows:

Eq. 1. $\quad \vec{E}_t = \vec{i}\,\vec{E}_t \cos\theta + \vec{k}\,\vec{E}_t \sin\theta$

Squaring both sides yields:

$$\vec{E}_t^{\,2} = (\vec{i}\,\vec{E}_t \cos\theta)^2 + (\vec{k}\,\vec{E}_t \sin\theta)^2$$

Taking the square root of both sides yields:

$$\vec{E}_t = \vec{E}_t (\cos^2\theta + \sin^2\theta)^{1/2}$$

But,

$$(cos^2\theta + sin^2\theta)^{1/2} = 1 = cos^2\theta + sin^2\theta$$

This identity has bothered the author, but $\vec{E_t}$ is a conservative vector and $\vec{i} \cdot \vec{i} = \vec{k} \cdot \vec{k} = 1$ seems to permit it. Theoretical assumptions may preclude this identity, but if the identity is allowed, the dual character of photon transfer follows:

$$\vec{E_t} = \vec{E_t}(cos^2\theta + sin^2\theta)^{1/2} = \vec{E_t}cos^2\theta + \vec{E_t}sin^2\theta = \vec{M_t}c^2$$

Allowing $\vec{E_t}$ to take on both "pure mass" and "pure energy" characteristics, the Dual Function Equation follows:

Eq. 2. $\vec{E_t} = \vec{M_t}c^2 cos^2\theta + \vec{E_f} sin^2\theta$

Where:

and: $\vec{M_t}c^2 cos^2\theta$ is the "pure mass" function vector in the *x* direction

$\vec{E_t} = \vec{E_f}$, or $\vec{E_f} sin^2\theta$ is the "pure energy" field vector in the *z* direction.

This same Dual Function Equation can be shown as a combination of two planar waves propagating along the *y* axis, as shown in *Figure 2* (page 3).

Before going on, the author would like to point out that this equation is inconsistent with Maxwell's equations because it allows only one set of values for the "pure mass" function vector and the "pure energy" field vector for any *(dy)* element along with *y* axis; i.e., the *(dy)* element will not experience a maximum $\vec{M_t}c^2 cos^2\theta$ at one time and a maximum $\vec{E_f} sin^2\theta$ a short while later because if that *(dy)* element were a receptive matter particle, the particle would be allowed to

2 – Addendum

experience an increase in energy = $2\vec{E_t} = \vec{M_t}c^2 + \vec{E_f}$. This rationale leads to the following postulation:

> "self–propagating waves should not be viewed as a wave–form sliding along the axis of propagation (as a wave traveling in a rope), but that a self–propagating wave vector set decays or vanishes to another vector set as the wave travels",

recalling that electromagnetic waves do not need a media to travel in as mechanical waves do. A somewhat different approach to media vs non–media for photon transfer will be brought out in Section IV of this paper.

Figure 2.

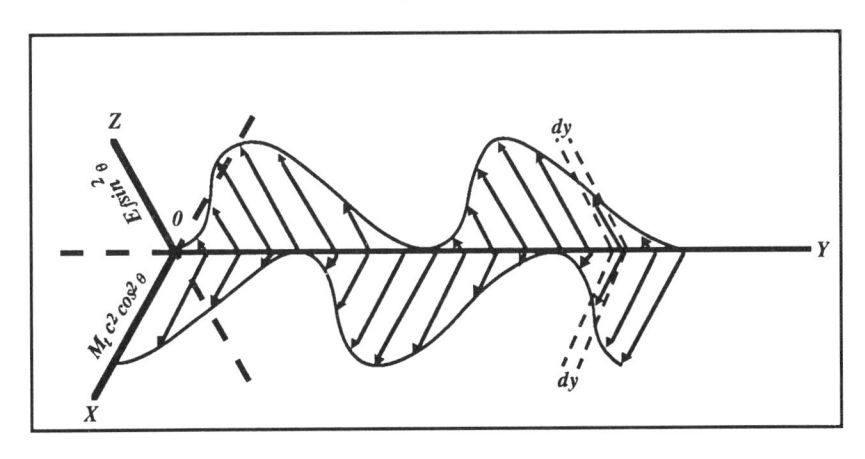

Figure 2, $\vec{E_t} = \vec{M_t}c^2\cos^2\theta + \vec{E_f}\sin^2\theta$ propagating along the y axis.

II. DERIVATION OF PHOTON MOMENTUM

Because momentum has been defined as a mass–motion equivalent, this portion of this paper will deal only with the "pure mass" vectors of the Dual Function Equation; however, there is strong belief that a redefinition of momentum is in order. Reason for a redefinition will be brought out in Section V, Reflection Phenomena, of this paper.

If a "pure mass" vector set is associated with photon transfer during a period of time, as shown in *Figure 3* (page 4), then momentum should be predictable by the Dual Function Equation.

Figure 3.

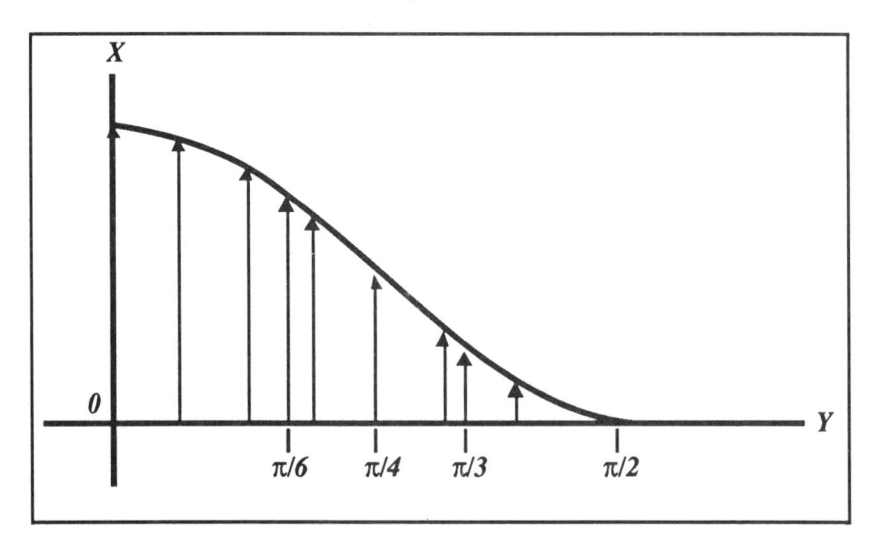

Figure 3., A "pure mass" vector set during the interval $0 \rightarrow \pi/2$

Referring to *Figure 3*, the "pure mass" function is given by:

Eq. 3. $\qquad f(M_p) = \vec{M_t} \cos^2\theta = \dfrac{\vec{E_t}}{c^2} \cos^2\theta$

Then the cumulative sum of the "pure mass" vectors during the interval $0 \rightarrow \pi/2$ is:

Eq. 4. $\qquad \displaystyle\int_0^{\pi/2} f(M_p) = \vec{M_t} \int_0^{\pi/2} \cos^2\theta\, d\theta = \dfrac{\vec{M_t}\, \pi/2}{2}$

The average "pure mass" vector during this interval is:

Eq. 5. $f(M_p) \, avg. = \dfrac{\vec{M_t} \, \pi/2}{2 \, \pi/2} = \dfrac{\vec{M_t}}{2}$

As could be deduced by symmetry.

Note: All mass vectors are traveling along the *y* axis at velocity c_m (the velocity of a photon in its media). Therefore, the average momentum (P_{avg}) is:

Eq. 6. $P_{avg.} = \dfrac{\vec{M_t} \, c_m}{2}$

Equation 6 is interpreted as the *expected* contribution of momentum by a single photon, in a beam of photons randomly out of phase but having the same frequency and direction, at any *(dy)* element. The expected momentum for the beam is:

Eq. 7. $p = \dfrac{n \vec{M_t} \, c_m}{2}$

Where:
n = number of photons

$\dfrac{\vec{M_t}}{2}$ = average mass vector of the beam

c_m = speed of light in its media

For any *(dy)* element individual photons do not have an average momentum, but have an instantaneous momentum of :

$p = M_t \cos^2\theta \, (c_m)$.

Replacing $\vec{M_t}$ by $\dfrac{\vec{E_t}}{c^2} = \dfrac{h\nu}{c^2}$ yields:

Addendum – 5

Eq. 8. $p = \dfrac{nh\nu c_m}{2c^2}$ for the beam

Eq. 9. $p = \dfrac{h\nu c_m \cos^2}{c^2}$ for the photon

Equations 8 and 9 indicate that momentum is directly proportional to ν and c_m. It is felt that momentum is directly proportional to ν but not to c_m. In other words, ν is a reflection of the quantity of E_t upon emission of the photon, but because a photon slows down for a period of time (in a pane of glass) does not necessarily mean that the photon cannot resume its former wave length and frequency upon emission from this media (glass). However, this does not preclude the photon from giving up energy to this media (glass), but if it does not, it will then resume its former wave length and frequency upon emission from such media. The author is lead to the postulation that $\nu_m c_m = $ a constant $= \nu c$ in free space.

Where:

$\nu_m = \nu c / c_m$

Then

Eq. 10. $p = \dfrac{nh\nu_m c_m}{2c^2}$ for the beam

Eq. 11. $p = \dfrac{h\nu_m c_m \cos^2\theta}{c^2}$ for the photon

ν_m is not the true frequency of the photon while it is in the media, for the true frequency in the media is ν multiplied by an amplitude factor c / c_m. This amplitude factor keeps the area under the $\cos^2\theta$ curve constant even though the wave length changes.

It follows then that as the photon's speed decreases, the mass and energy vectors increase in magnitude. This is in conflict with Einstein's *THEORY OF RELATIVITY*, ("that mass grows to infinity as a particle is accelerated to the speed of light.") because to conserve momentum, the mass function of a photon must increase as its velocity falls away from the speed of light.

III. DISCUSSION OF REFRACTION AND DIFFRACTION PHENOMENA

The author has not yet mathematically formulated the laws governing refraction and diffraction, but he feels fairly certain that the Dual Function Equation will predict these phenomena. These phenomena are believed to be gravitational responses between the photon's mass vectors and the mass which the photon either penetrates or passes close by.

A. Refraction

As a photon approaches a body of transparent mass, at some incident angle, the interaction of the associated gravity fields of its mass vectors with the transparent body's gravity field will cause the photon to bend towards the transparent object. Once inside the transparent object, the photon will have approximately the same amount of mass on all sides and will then proceed in a straight line. Some variance might be expected if the transparent body did not have a uniform sectional density due to non–homogeneity of lattice–mass, or irregularity in its geometric shape. In either case, the photon would be attracted to the center of the influencing mass and its path will bend slightly while inside the object. Upon leaving the transparent mass, the photon would again be bent toward the center of the influencing mass of the transparent object.

B. Diffraction

Diffraction is believed to be a similar gravitational interplay caused by photons traveling past an object. As a diffraction slit or aperture is narrowed, the gravitational field of the mass bounding the slit or aperture has less area to influence and one would expect a greater bending response from the photons, thus causing a wider diffraction pattern. An experiment seems to be in order here; if one were to create two diffraction slits of the same width but with different material densities (such as paper vs. platinum foil), would two different diffraction patterns result? If variances do occur, it may lead to further formulations of the Dual Function Equation. Also, if this is a gravitational phenomenon, the diffraction pattern should be dependent upon the intensity of the beam through the slit; i. e., the central band should increase in brightness faster than the second and higher order bands as the intensity of light passing through the slit increases. If diffraction was the result of a purely geometric configuration with no other physical laws governing, then a uniform increase in intensity for all bands might be expected.

IV. DISCUSSION OF MEDIA VS NON–MEDIA TRANSFER OF PHOTONS

It has long plagued physicists whether electromagnetic waves do or do not require a media for transfer. It can be said that there are media which affect electromagnetic wave transfer, for in transparent mass media, electromagnetic waves slow down. This fact alone is at least a start for determining media vs non–media transfer. The author feels that this change in velocity is also an interplay between the associated gravitational fields of the mass vectors of the photon and the gravitational fields of the transparent mass media.

It is postulated here that as the photon's mass vectors begin to grow, they are further accelerated due to the attraction of the mass surrounding them. This increase in acceleration causes the amplitude of the wave to increase as discussed in Section II, Photon Momentum. If the frequency, v, remains constant, then dM_p / d_y increases and if the E_t function of the photon is going to be conserved $\lambda \rightarrow (\lambda - d\lambda)$ such that $v(\lambda - d\lambda) = c_m$ and $c_m = v_c / v_m$. It is thought that the change in wave length is a "result" of the photon's vectors responding to their media and not the "cause" of an increase in amplitude in the wave function.

What then of free space? Well, free space, by definition, seems to be those portions of the Universe void of matter particles. One can evacuate a vessel but it's more difficult to remove all of the electromagnetic waves from inside the same. As postulated by the Dual Function Equation, if electromagnetic waves exist, then "pure mass" vectors exist also.

It is felt that electromagnetic waves *do* need a media for transfer, but that that media is and has been disguised as the electromagnetic waves that flood "free space"; i.e., electromagnetic waves need of themselves more electromagnetic waves for transfer. The Dual Function Equation predicts that during any cycle, the "average" $\vec{M_t}$ $c^2 cos^2 \theta$ = the "average" $\vec{E_f} sin^2 \theta = 1/2\ E_t$. A group of photons occupying the same spatial zone would interact with each other to form an "average" $\vec{E}_{t_e}$, given the subscript e for environmental average. This $\vec{E}_{t_e}$ may be viewed as a diffused, transparent resonant mass–energy body but more correctly it is the cumulative sum of all mass and energy vectors that will interact with a particular photon at a given spatial position. By interacting with this E_{t_e}, the photon's vectors also contribute to the E_{t_e} of surrounding photons. E_{t_e} is believed to be a steady state flux with 1/2 pure mass function characteristics and 1/2 pure energy characteristics for free space. It is postulated that as a photon's vector set begins to grow form zero, they are accelerated toward the E_{t_e}, upon passing the $\vec{E}_{t_e}$, (steady state average) the set's vectors will experience a retarding acceleration toward the E_{t_e}. This stored potential reaches maximum at $\vec{M}_t c^2 = \vec{E}_f = \vec{E}_t$. If this postulation is real, then the following conclusions could be drawn:

1. A photon will not be emitted or transmitted if its $\vec{E}_t$ is insufficient to interact in this way with the $\vec{E}_{t_e}$ surrounding the photon.

   a. Emission from a dense mass vector environment will probably begin as a growth of an $\vec{E}_f$ vector set.

   b. Emission from a dense energy vector environment will probably begin as a growth of a $\vec{M}_t c^2$ vector set.

2. The photon's media determines its velocity. All mass and energy vectors of an emitting photon will be *orthogonal* to its direction of propagation and the photon's transfer velocity is limited by the media's potential to retard or enhance growth of these *orthogonal* vectors.

3. The direction of propagation of the "media waves" need not determine the speed or direction of the photon; i.e., the "spatial average" is $\vec{E}_{t_e}$ in all directions. Polarization, electropotential fields, magnetic fields and gravity are forms of ordered mediae, and they will impose directional and velocity changes to the photon.

Addendum – 9

Detection of such mediae responses are evident in refraction and diffraction, and the author presumes that future formulations of physical law will bear these phenomena out as ordered photon response to its surrounding ordered mediae; otherwise, photons may alter the direction of their momentum without cause.

Detection of an "ether" as viewed by earlier physicists may prove futile because it is felt that in all types of media there exists mostly uniform homogeneity in all directions; furthermore, the "ether" is always changing, but the "average" state remains constant. When a photon is subjected to variation of media types over its path length, then ethereal effects will become apparent. This response caused by ordered mediae is viewed complete from emission (least ordered mediae response) to absorption (most ordered mediae response), but its detectable limit may prove in between these two extremes.

V. DISCUSSION OF REFLECTION PHENOMENA

Although reflection is one of the more rudimentary studies of photon transfer, it has never been clearly defined as to what physical law(s) cause this photon response. Perhaps there have been theoretical dissertations on the subject of photons bouncing off objects, but the author has not had the opportunity to peruse any.

The Dual Function Equation predicts that reflection only occurs as a boundary phenomenon. A photon will not reflect if a "pure energy" vector is present within the *(dy)* element where motion in the *y* direction ceases. The photon must be totally "pure mass" to reflect *($\vec{M_t}$ $cos^2\theta = \dfrac{\vec{E_t}}{c^2}$, and $\vec{E_f} sin^2\theta = 0$)*. This is the same type of boundary condition described in Schrödinger's wave mechanics.

Derivation:

1. Momentum = $p = \dfrac{U}{c} = \dfrac{E}{c} = \dfrac{h\nu}{c}$ total absorption

$$p = \dfrac{2h\nu}{c}$$ total reflection

2. Dual Function Equation = $p = \dfrac{h v_m c_m \cos^2\theta}{c^2}$ for a photon (Eq. 11.)

$v_m c_m = vc$ (author's definition)

$p = \dfrac{hvc \cos^2\theta}{c^2}$

$p = \dfrac{hv}{c} \cos^2\theta$ @ $\theta = n\pi$, $p \to \dfrac{hv}{c}$ (total absorption)

$p = \dfrac{2h \cos^2}{c}$ @ $\theta = n\pi$, $p \to \dfrac{2hv}{c}$ (total reflection)

$n = 0,1,2,3...$ and $E_f \sin^2\theta = 0$

Note that the author has shown equivalence using only the mass function of the Dual Function Equation with the aid of Schrödinger's boundary condition. This was done because momentum is defined as a mass–motion equivalent. The author feels that reflection can also occur as an elastic collision between two "pure energy" states or two simultaneously "pure mass" – "pure energy" states. The author feels that Schrödinger's boundary conditions perhaps prevail for abrupt changes in photon direction, but there seems a need for gradual changes in the direction of photon momentums such that bending of photon paths may occur uniformly. Because mediae of "pure energy" vectors may cause such bending, momentum is believed to be an energy–motion equivalent also.

[Based upon momentum being a mass–motion equivalent, the "average" momentum discussed earlier in Section II of this paper has no meaning for reflected and absorbed photon, if boundary conditions prevail. This "average" momentum is a position vector. For total reflection to occur, the penetrable *(dy)* element on the mirror's surface must be at least *1/2 λmax* from the *(y)* position of the first reflected photon, to the *(y + dy)* position of the last reflected photon — for noncoherent, nonmonochromatic light. This distance allows all photons to grow into a "pure mass" state.]

It is thought that these reemissions are both "pure mass" and "pure energy" elastic collisions. The reemitted photon will begin along the path of reflection as a growth of the opposite "pure" vector it collided with; thus, a photon obeying the law $E_t = M_t c^2 \cos^2\theta + E_f \sin^2\theta$ will upon reflection obey the law $E_t = E_f \cos^2\theta + M_t c^2 \sin^2\theta$. The time required for emission or reemission is a moot point, but the time to insure emission or reemission is $T = (1/4\lambda / c_m) 2$, assuming a velocity change of $0 \to c_m$.

Reflection is defined here as, "the phenomenon which results when a photon is unable to effect a change in the Entropic or Inertial Potentials of an object's media." Reversal of the photon's vectors takes place and the object as a whole suffers an increase in Entropy. E_t of the photon has insufficient energy and/or freedom to interact with E_{t_e} of its mirror. (Entropic and Inertial Potentials are discussed in Section VI.)

VI. DISCUSSION OF ENTROPY AND INERTIA

Introduction

On a more personal note, the author would like to express that of all of the discussions in this concept, this particular one consumed more time and thought processes than any other. For unless one can show basic definitive continuity from the tiniest quantum system to the largest cosmological event, he has failed to bring the true scope of resonant mass–energy systems to light. The author does not ascribe to spontaneous or random events occurring without cause, and feels to do so is an admittance of lack of understanding. (He sometimes alludes to present–day man as his not–so–distant ancestor, the Caveman: who perhaps watched fruit from the tree of knowledge fall *randomly* to the ground and then one day an unseen hand shook the tree and it gave forth fruit abundantly. After eating his fill, he rested a time–then hunger set in, and he thought, I shall use this knowledge and shake the tree for myself. In so doing, he bruised the tree and denied the gentle unseen hand that fed him.)

Along similar lines of thought, the author does not ascribe to the Second Law of Thermodynamics or to time being relative such that mass zones can be constructed of simultaneous fragments having different times. Rather, he ascribes to an inertial reference frame somewhere in space from which constancy abounds in the decay and regeneration of mass–energy systems; thus, time is the constant by which variances in the rapidity of these dynamic processes can be measured. This perspective is taken so that fictitious vectors will not be introduced. Hopefully, adherence to this perspective will not lead to erroneous conclusions.

Until the author stumbled upon the Dual Function Equation, electromagnetic energy was thought to be a different substance than matter. It is true that "pure energy" (not being defined here) is a different substance than "pure mass" (also not being defined here) but classical physics has ascribed that "pure energy" = electromagnetic energy and "pure mass" = amount of matter. It is postulated here that electromagnetic energy systems are only different dynamic configurations of matter systems which are both comprised of "pure energy" and "pure mass" vectors dynamically decaying and regenerating. Thus, if we view the Dual Function Equation, $E_t = Mc^2 cos^2\theta + E_f sin^2\theta$, from classic and relativistic perspectives, the systemizing element, E_t, is erroneous and the equation degenerated into Einstein's conversion equation $E = Mc^2$. Although Einstein's conversion equation has the proper equivalence it does not show identity of electromagnetic systems to matter systems; therefore, one fostered in classic and relativistic perspectives would conclude that these systems are two different substances; namely, "pure energy" and "pure mass". However, if these systems are viewed from the Dual Function Equation perspective, both equivalence and identity can be inferred.

Discussion

In order to conceptualize Entropy and Inertia based on a Dual Function perspective, the author chooses to define two extremes of resonant mass–energy systems and four terms. These extremes do not preclude intermediate combinations; for the spectrum of resonant mass–energy systems would appear to be full from the "pure mass" state to the "pure energy" state.

The extreme systems are:

1. *Linear Motion Systems* – Where *orthogonal* growth and decay of "pure mass" and "pure energy" vectors interacting cause and stimulate linear self–propagations (electromagnetic waves). (See *Figure 2.* on page 3)

2. *Inertial or Satellite Systems* – Where growth and decay in inertial "pure mass" and "pure energy" vectors cause and are caused by growth and decay of *orthogonal* (orbital) motion of associated inertial systems (Universal Matter Systems). (See *Figure 1.* on page 1)

The four terms are:

1. *Entropy* – The quantitative measure of Linear Motion Systems in a spatial zone. (More commonly referred to as the measure of disorder in a spatial zone.) Units are mass–length^2 / time^2.

2. *Entropic Potential* – The potential for a growth in Entropy. Entropic Potential is inversely related to Entropy; i.e., as Entropy increases, the Entropic Potential decreases. Unitless with a range between– $-1 \leq ep \leq 1$.

3. *Inertia* – The quantitative measure of inertial Entropy in a spatial zone. (It may be referred to as stationary electromagnetic energy or mass function.) The important concept is that Inertia not only refers to motionless mass function, but also to the occupation of a zone of space by quasi–stable but potential Entropy, as defined above.

4. *Inertial Potential* – The potential for a growth in Inertia. Inertial Potential is inversely related to Inertia with unitless dimensions and a range of $-1 \leq ip \leq 1$. This term should not be confused with gravimetric potential which is part of the mass function. The gravimetric potential is inherent with the amount of mass function present, but Inertial Potential incorporates both the mass function's gravimetric potential and the mass function's environment.

In the classical perspective, it can be said that the Universe is decaying from "B", Inertial or Satellite Systems (classically known as matter) to "A", Linear Motion Systems (classically known as energy). Entropy is the quantitative measure of this decaying, or disordering, process. The Second Law of Thermodynamics implies that the Universe is destined to the highest degree of disorder; i.e., all matter is destined to decay into energy and shall not return to matter. The author does not agree with this concept for many reasons, and he feels that this conclusion has been drawn because electromagnetic energy (thought to be "pure energy") is considered a different substance than matter, rather than a different dynamic configuration of the same elements. The author is not questioning the obvious ramifications of the Second Law of Thermodynamics such as heat always flows from a hot to a cold object; however, the farther reaching concept that there is a built in potential for disorder and the Universe is responding to it is thought wrong. One of the greatest testimonies against such a hypothesis comes from nature itself; i.e., nature always expends the least amount of energy or matter to seek equilibrium and if for some reason it over expends itself, it will rebound in such a manner as to conserve both energy and matter (such is the case of Brownian Motion). The author recognizes that the Universe perhaps is in a state of matter (Inertia) decay, but he also recognizes a built in policing system, one referred to as Inertial Potential.

If one were to look for such a policing agent in action, what would he look for? In the classical perspective, he would probably look for matter growth at the expense of electromagnetic energy. Such phenomena can be created in a laboratory but it is a rare case to see this happen frequently in nature on a macro level. On a quantum level, however, there is at least strong indication that electromagnetic waves have a built in tendency for creation or recreation of matter. This indication comes to us in the form of refraction and diffraction. Under normal conditions, light is always bent in a manner such that it appears attracted to matter. If both electromagnetic energy and matter had similar potential for randomness (without any policing agent) then one might expect the beam of light to be bent away from the matter, similar to a beam of electrons passing by a highly negative charged sphere.

Another example of this policing agent on a macro level is our own mother Earth. Daily, the Earth pulls in thousands of tons of meteorites randomly wandering through the heavens. The capturing of matter in a more random mode and incorporating it into one larger body is an ordering process. Note that if the Earth were suddenly incorporated into the Sun, its mass function probably would not have this potential for ordering. Rather, like a green–growing twig placed on a fire, the Earth potential for life would die, equilibrium would be established and its matter (inertia) would begin to decay, showing a negative Inertial Potential, or positive Entropic Potential.

A number of conclusions can be drawn from these two descriptive illustrations, some of which are listed below:

1. Laboratory results have proven that if high energy electromagnetic waves are forced into the same spatial position, a mass particle will form. As a result of this *net-cancellation of energy*, subsequent lesser electromagnetic waves are given off *orthogonal* to the incident waves. This shows possibility of creating Inertia from randomness. Presumably, as more

randomness occurs (as more electromagnetic waves flood space) this event becomes more probable.

2. Ordering can also occur by the accumulation of Inertia; thus, an over energized electron will emit electromagnetic energy and cascade into a vacant lower energy level to preserve Inertia (preserve the stability of its atom's nucleus).

3. The mass function has a built in policing parameter (gravimetric potential) which in and of itself is not the sole determining factor for restoration of Inertia; i.e., the environment of a matter particle also plays a role in determining whether Inertial Potential is positive or negative. As in the self-propagation of linear motion systems described in Section IV, MEDIA VS NON-MEDIA TRANSFER OF PHOTONS, so, too, there is a need for an E_{t_e} for macro systems. At this time, redefinition of E_{t_e} seems to be in order; i.e., E_{t_e} is the cumulative sum of *all* "pure mass" – "pure energy" vector sets that will interact with a particular "pure mass" – "pure energy" vector set irregardless of what resonant mass-energy system possesses these vector sets. This statement not only infers decay and regeneration of vector sets at and for the decay and regeneration of other interacting vector sets but also infers that "pure mass" and "pure energy" vectors always exist in simultaneous pairs (either of which may be zero for an instantaneous period of time).

4. The process of photosynthesis is also in violation of the Second Law of Thermodynamics, for this process converts random electromagnetic energy into Inertia. The unleashing of stored fossil fuel energy to create systems out of equilibrium with their environment and the monitoring of subsequent energy flows is not a test for decay or randomness but only the returning of Inertia from whence it came. The point here is that restoration of Inertia had already taken place prior to experimentation, and that the Earth's state of Inertia (in regard to fossil fuels) is too large for the experimental environments beset upon these fuels. Our ancient tree of knowledge did not deny the gentle unseen hand that fed it.

Summary

In summary, we have the following:

1. Possible affinity between Inertial and Linear Motion Systems (refraction, diffraction);

2. Possible restoration of Inertia from *net-cancellation* of Linear Motion Systems (sub–atomic level);

3. Preservation of Inertia in the nucleus at the expense of Linear Motion Systems being emitted (atomic level);

4. Restoration of Inertia by photosynthesis (chemical level); and,

5. Restoration of Inertia by accumulation of matter (macro level).

With the above taken into consideration, there seems to be a need for establishing: a) an opposite to Entropy – (Inertia), b) a potential for Inertia – (Inertial Potential), and c) an environmental parameter partially responsible for Entropy, Entropic Potential, Inertia and Inertial Potential – (E_{t_e}) which is both initiator and result of the terms above.

It is at this time, that the author postulates that the Dual Function Equation coupled with E_{t_e} and the terms at the beginning of this Section will apply to all resonant "pure mass" – "pure energy" systems. Perhaps this Dual Function Equation has little practical application at this time in history, but the theoretical implications of it seem sound.

Discussion on Disorder

One of the more classical discussions on disorder involves the irreversible–isothermal–adiabatic expansion of a gas from one vessel (A) into an evacuated vessel (B). If we let the number of gas molecules = $2n$, then upon expansion there will be (n) molecules in each vessel. The probability of finding a gas molecule in a given fixed volume increment (dv) will be higher before expansion than after, and Entropy, being equated as an inverse function of this probability, is higher after expansion. Also it is a stronger statement to say that there are $(2n)$ molecules in vessel (A) than to say there are $(2n)$ molecules in vessels (A) and (B) combined.

Addendum – 17

If the process described above is applicable to the presumable Entropy growth occurring throughout the Universe, there are two fundamental questions to be answered. They are:

1. Because we have created an artificial freedom for expansion, does this freedom exist throughout the Universe? (To allow this freedom would seem the weakest statement of all.) And,

2. Conversely, is the size of the Universe fixed? The latter is a weak statement also, but based on the previously mentioned policing agents, we might conclude this. If we fix the size of the Universe, we are then faced with the application of our artificial experiment to the real world. In order to do this, we would have to expand our sampling space to *2(dv)* as the gas expands, thus there would be no increase in Entropy.

If we did not choose to fix a finite size to the Universe, then we could allow the gas to expand infinitely. We could accomplish this by attaching *(n–1)* evacuated vessels of equal size to both (A) and (B). After opening these vessels, simultaneously, we would expect one molecule to travel to each vessel leaving one behind in both (A) and (B). The probability of this event occurring would be rather low and yet the system has the freedom to allow this. Also, in the event this did happen, perhaps the strongest of statements that could be made about any of the systems above is that there is one molecule in each of the vessels. (Granted the probability of finding one molecule in our original *(dv)* increment would be the least, but somewhat like a Klein bottle, our logic has turned in on itself for now we can say that there is one molecule in region 1, in region 2, in region 3, etc...) Lastly, even as boundless as our logic might presume the Universe to be, there would seem a high probability of two or more molecules approaching one another at some future time; therefore, we might expect two or more molecules to return to vessels (A) and (B) simultaneously. If a probability be assigned to this last phenomenon, then it will also limit Entropy and the size of the Universe and we must turn to assumption 2. above.

Another possibility is that all of the molecules remain in each of their respective regions, but this is a fixed configuration like unto Inertia. This is probably a reiteration of our Klein bottle, or when Entropy quantitively increases the quality of segregation also increases.

This seems in part counter logic to randomness, for randomness would imply the inability to *earmark* a separate entity.

With these thoughts in mind, the author finds the classical definition of Entropy, as a measure of disorder, nebulous. For to measure disorder, we have to mathematically contain both the starting products and the results which is an enigma; for we try to contain an uncontainable phenomena in a Universe that is boundless. This classical concept also leads us to the somewhat juvenile but unanswerable question, if the highest order of randomness is the order of the day, then why didn't the Universe start that way? It would seem that the only answer to this is that the Universe was first created, and the laws which govern the Universe's behavior came after. The two events could not happen simultaneously. Such logic would lead us to believe that all phenomena are free from and produce physical law. This would also imply that the Universe is a perpetual motion machine without internal evidence of this phenomena.

Further Discussion on Disorder

The Second Law of Thermodynamics implies that it always takes more energy to restore order than to cause randomness. A simple *irreversible* process will prove otherwise. It is referred to as irreversible because the vector *time* is not reversible. Take one three foot square table and grid it off so that there will be nine equal one–foot squares. Place nine lumps of coal in the center square. If an experimenter then begins to pick up eight of these coal lumps, one at a time, placing them in the center of each of the vacant one–foot squares, he will do more work than if he reverses the process after distributing the coal lumps. This is true because in disordering the original pile, he has to overcome the gravitational forces that tend to hold the lumps together. In the reverse process, these gravitational forces reduce the experimenter's efforts. Note, too, that the center of gravity will not change location or magnitude whether the coal lumps be ordered or disordered (in regards to some object lying in the same plane at the other end of the room). If the disordered array represents an increase in Entropy, then the disordered array's gravitational field should be less (with regard to the far object) because its arrangement corresponds to a loss in mass function equal to dQ (representing the energy expended to overcome these gravitational forces). Furthermore, this lost mass function (if real) can be restored with less expense of energy than it took to destroy it per classical theory. The author does not feel that there is a loss in mass function but a loss in Inertia with the disordered state. When the coal lumps were all in one pile, there was a greater *net-*

cancellation of gravitational field energy between the lumps because they were closer together. When the coal lumps are farther apart, there is less *net-cancellation* of gravitational field energy and this lost gravitational field energy *net-cancellation* is manifested as an increase in mass function of each coal lump; otherwise, the experimenter has exchanged a larger *net-cancellation* for a smaller *net-cancellation* and annihilated energy. When the coal lumps were close together, these higher energy *net-cancellations* represented an increase in Inertia, but as the coal lumps are disordered there is also an equal and opposite *net-cancellation* of energy equal to dQ; thus, upon disordering, the system retains the same mass function Inertia cumulative quantity with an increase in Inertial Potential. (Refer to original concept where "pure mass" is a *net-cancellation* of "pure energy".)

The object on the far side of the room (in regard to its mass and gravitational field energy) may be considered as E_{t_e}. Its gravitational energy (upon disordering the coal lumps) no longer acts towards the center of the table but acts radially inward towards itself from each arrayed coal lump. Because of this, the experimenter has also altered E_{t_e}, for now there are gravitational field energy *net-cancellations* where they did not exist previously. This concept of fixed cumulative amounts of mass function – Inertia and the lack of field energies where there is nothing to interact or provide *net-cancellation*, demands a fixed size to the Universe with a fixed reference frame. The central square on the table is the fixed reference frame for the system because there is a fixed mass function – Inertia quantity about it. The fixing of the Universe is necessary so that interaction of vectors can occur, otherwise, "pure energy" – "pure mass" vectors can exist without interaction. The concept of a vector entails magnitude and direction—without interaction a vector has neither. (Fluctuation in the size of the Universe is not precluded with this concept, but repetition of fluctuation is demanded.)

This concept may also be applied if the coal is ignited and subsequently disordered. The problem here is that the experimenter may be unable to contain the disorder because he has converted an Inertial System to a Linear Motion System (a self–sustaining system which interacts with its environment until it reaches a position of high Inertial Potential and is converted totally, or in part, back into an Inertial System. (The completeness of restoration of Inertia will be related to E_{t_e} upon conversion.)

In regards to the mass growth of the coal lumps upon disordering, an experiment that may shed some light on the subject might be: Does one gram of magnetized iron contain the same number of molecules as one gram of nonmagnetized iron? A lack of clairvoyance is stressed here, because it is not clear to the author if the alignment of dipoles are not also *net-cancellations*, and, therefore, the number of molecules would be the same for both states, or opposite from the prediction that follows. If it is experimentally feasible to show a difference in molecular count, the author would predict that the magnetized iron would have more molecules per gram than the non–magnetized iron. The reason being, that the non–magnetized iron may have more internal *net-cancellations* (a higher state of Inertia) than the magnetized iron which has more internal field energy. The opposite view may be taken here in that because the crystal lattice constrains the randomness of oscillations of the dipoles, it forces a higher state of Inertia upon them, and they would thereby have more mass per dipole.

The Need for Motion

Motion is necessary in both the Linear Motion System fraction and the Inertial System fraction of the Universe. It is both result and cause of each E_t fraction interacting with E_{t_e}. In the abstract sense, motion may be considered evidence of an interaction between E_t and E_{t_e}.

This latter statement manifests itself when considering state variables derived from nonstate variables. The variations in motions, i.e., the variations in paths between two fixed states of entropy by nonstate variables (per classical theory) is of noneffect in the Dual Function perspective because each path variation will result in a different $E_t - E_{t_e}$ combination. dQ & dW are state variables in the Dual Function perspective and as such $dS = dQ / T$ expresses both equivalence and identity. This is true because dQ and dW are relationships of simultaneous "pure mass" – "pure energy" vector sets, rather than just "pure energy".

Variations in the Mass Function Due to Motion

Mass Decay:

In any perspective of resonant mass–energy relationships we must come to the conclusion that there are two ways in which gradients of potential Entropy or Inertia may be subsequently relieved;

1. by motion or cessation of it,

2. by direct conversion from a Linear Motion System to a Universal Matter System, or vice versa, and

3. systems expressing relief by both 1 & 2 occurring simultaneously.

After recognizing these two types of relieving processes, if we then look at nature, we probably would come to the conclusion: that the larger a body's Inertia, the slower or less complex the body's motion; and the slower or less complex the body's motion, the greater the Entropic Potential for *direct* conversion of the body's Inertia into Entropy.

Let us investigate the motion of a $-\beta$ particle resonating in a neutron of a U_{239} nucleus. It is not just resonating as part of this neutron, but rather its motion is much more complex. It is resonating as part of a neutron, which is part of a nucleus, which interacts with an electron cloud to form a quasi-stable atom, which resonates in a crystal lattice, which spins about the axis of the earth, which orbits about the sun, which spirals through space, etc. Because each subsequently larger mass entity predominantly influences the motion of each lesser mass entity, we might conclude that the $-\beta$ particle is where it is because of these larger mass entities and their motions.

In a similar manner, we might explain each *direct* exchange from Inertia to Entropy as a consequence of Entropic Potential and an E_{t_e} favorable for relieving this Entropic Potential gradient. Thus, we would expect zones of great regional Inertia to have a greater tendency to radiate Linear Motion Systems. This is the conceptual approach to the critical mass/packing fraction relationship already in employment by physicists to explain nuclear radiation. The same concept may be extended to macro mass entities such as stars. It is of some importance to note that because critical mass zones are not only subject to physical quantity of mass but also to E_{t_e}. Under certain E_{t_e} conditions, supercritical masses could exist and others could become critical, even though they contained a subcritical amount of mass under normal E_{t_e} conditions.

In short, both nuclear and solar radiation phenomena are not the result of random processes, but are the consequences of high Entropic Potential gradients and an E_{t_e} compatible with relieving these high gradients. One might look at the natural decay of our $U_{239} \xrightarrow{-\beta} N_{p239}$ as a relieving process for:

1. Gradients set up by irregularities within the normal resonant patterns of nucleus and electron cloud motions;

2. Gradients set up by the Earth slowing down, i.e., the Earth is gaining Inertia; and

3. Gradients set up by the Earth accumulating mass—or accumulating Inertia by ordering heavenly bodies.

Other forms of Inertia loss are:

1. Linear Motion System evolution due to motions forced upon Inertial masses.

2. The acceleration of a photon were $v_m \rightarrow v$ as discussed on pages 5 & 6.

Mass Growth:

Einstein predicted that if a mass particle is accelerated toward the velocity of light, c, the particle will increase in mass. The author agrees with this relativistic concept but perhaps for a different reason.

When the mass particle is at rest, it has a more positive Entropic Potential for mass decay because of normal gradients set up within our Inertia – Entropy fractions of the Universe.

As the particle begins to accelerate, its motion relieves this positive Entropic Potential, and as it approaches the velocity of light, the simultaneous "pure mass" – "pure energy" *orthogonal* vector sets making up this mass particle become less and less different (distinguishable) from the accelerator's Linear Motion System "pure mass" – "pure energy" vector sets. The *net-cancellations* set up between the mass particle's vector sets (E_t) and the resisting accelerator's Linear Motion System's vector sets (E_{t_e}) gives rise to Inertial increasement. If the accelerator's vectors are removed, i.e., if (E_{t_e}) is decreased, then this increased Inertia will dissipate as Linear Motion Systems until E_t reaches equilibrium with its new E_{t_e}. The author feels that this likeness in vector character must be brought about for unless vectors be of like character they cannot cancel one another.

Although it may seem confusing, the same thing happens when a photon passes by or through an object of mass. There is a mass growth (or its equivalent) as the photon is attracted by the mass and is subsequently slowed down: thus, Einstein's hypothesis of mass growth with acceleration does not apply to Linear Motion Systems, but to Universal Matter Systems only. The *net-cancellations* of a photon's gravitational field with the surrounding mass's gravitational field must result in an increase in mass function (Inertia). This increase in mass function (Inertial) should directly correlate to the increase in frequency amplitude, i. e., $v \to v_m$ as discussed on pages 5 & 6. If this were not so, then entropy per classical theory would be the order of the day and photons would be unusually large. The Relativistic Concept that a photon has zero rest mass is thought to be a cop-out because it allows a mass equivalent substance (energy) to be transported from point A to point B, and yet a loss of mass occurs at point A. If photons *do* have a rest mass, then they should obey Einstein's mass growth hypothesis; we can thank our benevolent stars that photons are disobedient.

A question arises here: If there is an increase in mass function for a photon as it slows down, does $E_f \sin^2 \theta$ also increase? The author speculates that it need not and the corresponding Dual Function Equation would be of $E_t + E_{t\ cancellation} = A\ Mc^2\ \cos^2 \theta + B\ E_f \sin^2 \theta$, where $A > B$.

The circle in *Figure 1* (page 1) becomes an ellipse elongated along the X axis. In an environment with less than $1/2\ Mc^2 \cos^2 \theta$ character, the ellipse should be elongated along the Z axis or $A < B$.

The Velocity of Light

In order to discuss the velocity of photon transfer, we must first define the nature of light. An important question comes into focus here and that is: Does a photon have substance? (Either mass or energy.) or, is it a concept? (A phenomena which can be observed but has no mass equivalent quantifying substances only time and /or location coupled with a noticeable character—examples: wave front phenomena, thoughts, etc.)

The author contends that photons are of substance, for if they are not of substance, they would not stimulate things of substance such as chlorophyll, photo electric cells, nerve cell electronics, etc. The reason why it is felt that this is such an important question is because nothing of substance can reverse its direction without going through a state of zero velocity, or in other words, the speed of photon transfer is not a constant but a variable. Recalling the author's postulation on pages 6 & 9 that $v_m c_m = vc$, we can obtain an equivalence with the Dual Function Equation even though the speed of light varies, but with the Relativistic Theory, $E = Mc^2$, the speed of light must remain constant in order to preserve equivalence. The erroneous Relativistic result is that the speed of photon transfer remains constant, while time is warped. The author contends that time remains constant while the velocity of photon transfer varies.

Let us examine a photon travelling toward a large body of mass, such as a planet. The Dual Function perspective relates its transfer velocity (c) is due to growth and decay of vectors *orthogonal* to the line of propagation. If the photon did in fact go through a "pure mass" phase every 1/2 cycle, then we should expect the photon to speed up because it would be attracted by the interaction of its gravitational forces and the gravitational forces of the large body. In effect, the result would be an increase in velocity to *(c + Δc)* with a continuance of the same frequence. (In other words, if the photon were observed on the planet by an observer, it would appear that the photon had suffered a "red shift".) If, however, the photon suffered a pure elastic collision and was reflected, it would rebound with the velocity - *(c + Δc)* and by the time it reached its point of first observation, it would be transferring at velocity *c* in the opposite direction. If the distance between the point of first observation and the planet is represented by "d", then the time for the photon to go and return is not $T = 2d / c$ but $T = 2d / (C + \frac{\Delta c}{2})$ assuming constant gravitational acceleration.

Note: Photons can also be deceleration in the same way. A photon travelling from our sun to each of its respective planets suffers gravitational deceleration by the sun and acceleration by the receiving planets. This could, in effect, make "red shifts" of intra–solar–system photons less pronounced than those observed from other suns.

The author recognizes *apparent* "red shifts" due to emission of photons from bodies travelling at rapid velocities; however, the author has not recognized this phenomena as a doppler effect. Unless we relate this frequency shift to the initial development–time to insure transmission, $T = (1/4\lambda / c_m) 2$ (See page 11) there is no mechanism for expecting a doppler effect from photons. The reason why the author states this is because photon transfers are independent of their source (per Relativistic Theory) whereas mechanical waves depend upon a continuous source of energy transfer. In other words, the relative rapidity at which two or more photons escape from their source can in no way affect each other's frequency. Photons are separate entities with built in frequencies (per Relativistic Theory).

The author feels that a photon's frequency and wave length are both dependent on variations in the photon's E_{t_e}.

Thus, if we couple this "red shift" phenomena with the ellipse speculation found at the end of *Mass Growth* (page 24), we have three dimensional elasticity for any given E_t in response to any E_{t_e}.

Gravity

There is a noticeable difference in freedom between Linear Motion Systems and Universal Matter Systems. Linear Motion Systems would seem to have the ability to flood space more homogeneously, while Universal Matter Systems are locked into quasi–fixed inertial zones.

The author earlier stated that "pure mass" is a vector state where no energy exists. This was done to introduce the concept of resonant mass–energy systems; however, the fact is that every time a "pure mass" vector occurs, a "pure energy" field exists. It is known as gravity. One is hard pressed not to define gravity as part of the "pure mass" vector itself, but the author feels that it is not. Rather, it is a potential derived by position. If we define mass as a *net-cancellation* of energy, which by its Inertia occupies a zone of universal space, this zone will not be in equilibrium with the homogeneous spatial Linear Motion System surrounding it. This mass zone has the greatest potential to become homogeneous with the surrounding space system by cancelling itself with another non–homogeneous mass zone; thus, there is an attraction of these two mass zones (gravity). The response may result in orbital motion or combination of the two mass zones creating a larger mass zone. The acceleration of growth of a mass zone is dependent on the Entropic Potential vs the Inertial Potential whether that mass zone exists in a photon or in an inertial system. Both Entropic Potential and Inertial Potential are viewed as self–destructing potentials, for they create an environment which has greater potential to reverse itself.

Second Law of Thermodynamics

The author would like to conclude this section with a redefinition of the Second Law of Thermodynamics:

In any action or reaction between the Linear Motion System Fraction and the universal Matter System Fraction of the Universe, each fraction will expend as little of itself as necessary to reach an equilibrium position which in itself is constantly oscillating.

VII. DISCUSSION OF THE *THEORY OF RELATIVITY*

The author feels that the *Theory of Relativity* shows equivalence but not identity. It is an extension of centrifugal force; i.e., it is a combination of real and fictitious equivalent vectors used to change a problem from an inertial frame of reference to a dynamic frame of reference. As a man speeding around a corner feels that he is being pulled radially outward, so a man travelling at the speed of light may feel time ceases to exist. It is felt that time, by definition, is a universal constant by which we measure the relative motion of phenomena about us. The rapidity of these motions are predictable by the Entropic and Inertial Potentials and the freedom to release these potentials. If time ceases to exist at the speed of light, then how can one discuss variances in a photon's relative position, for these variances consume no time?

VIII. CONCLUSION

The Dual Function Equation is in its rudimentary form. Quantitatively, it is believed to be correct. Further identification will be brought about by breaking the "pure energy" vectors into electromotive and magnetic field vectors. These vectors are thought to be at $\pi/4$ and $3/4\pi$, negative values are presumed possible because the Dual Function Equation deals with $cos^2\theta$ and $sin^2\theta$. If one reduces the equation to $cos\theta$ and $sin\theta$ then $E_f / \sqrt{2}$ is the expected maximum magnitude for the electromotive and magnetic field vectors.

The mass function may also be broken down into orthogonal mass and gravimetric potential vectors. Note that c^2 has the correct units for gravimetric potential, $distance^2 / time^2$. If these two vectors (mass and gravimetric potential) are orthogonal, it would seem likely that angular rotation of the system will occur if they are not equivalent in magnitude. Planar waves can be developed by the angle between the mass vector and gravimetric potential vector to be less than orthogonal. Figure 4 indicates one possibility of the photon resonant mass–energy system:

Figure 4.

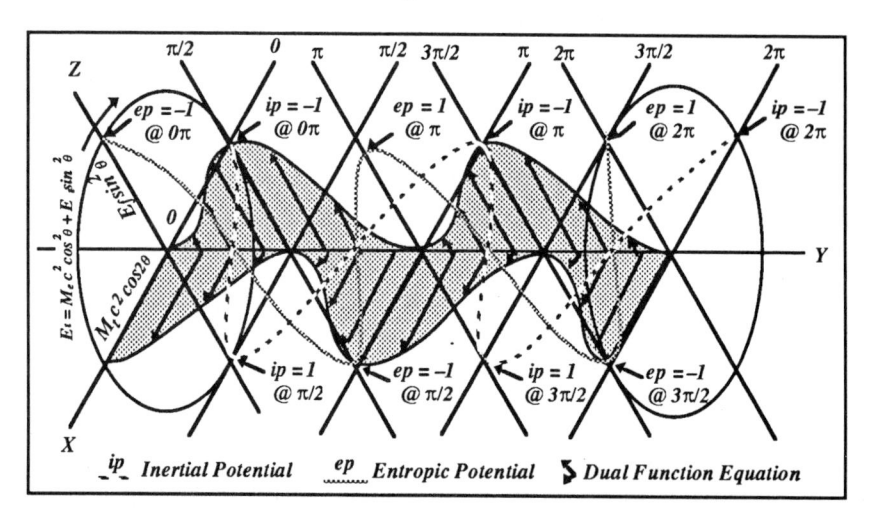

[Figure showing wave interference patterns with labeled axes X, Y, Z. Key markings include: $ep = -1$ @ 0π, $ip = -1$ @ 0π, $ep = 1$ @ π, $ip = -1$ @ π, $ep = 1$ @ 2π, $ip = -1$ @ 2π, $ip = 1$ @ $\pi/2$, $ep = -1$ @ $\pi/2$, $ip = 1$ @ $3\pi/2$, $ep = -1$ @ $3\pi/2$. Axis labels: $Et = M_t c^2 \cos^2\theta + E \sin^2\theta$, $E \sin\theta$, $M_t c^2 \cos 2\theta$. Legend: ip Inertial Potential, ep Entropic Potential, Dual Function Equation.]

Figure 4, One interpretation of the Dual Function Equation with time.

The author in his present academic state does not know how many of these concepts have been formulated previously, but at present, he offers the Dual Function Equation in the hopes of gaining a closer look at the profoundness of the Creator. As a concept, it may help to understand the following:

1. Self–propagation of Inertial and Linear Motion Resonant Mass–Energy Systems.

2. To infer identity between these two systems.

3. To understand refraction, diffraction and reflection of Linear Motion Systems (upon further formulation).

4. To define Entropic and Inertial Potentials.

5. To identify time as a universal constant.

6. To define momentum equivalent as both a mass–motion and energy motion equivalent.

7. To reintroduce an inertial reference frame back into physical law.

With much reluctance, the author has tried to abate the Entropic Potential to seek out and understand physical law, but Entropy prevailed. For the Dual Function Equation predicts, to know "pure knowledge" is to be in the highest state of potential for "pure ignorance".

I thank you for your time.

Notes

Notes